Famous Horses at War

American troops (and their cooks!) pay tribute to the 8 million horses and mules killed in the First World War.

Famous Horses at War

A Soldier's Mount Throughout History

M. J. Trow

Pen & Sword
MILITARY

AN IMPRINT OF PEN & SWORD BOOKS LTD.
YORKSHIRE – PHILADELPHIA

First published in Great Britain in 2022 by
Pen & Sword Military
An imprint of
Pen & Sword Books Ltd
Yorkshire - Philadelphia

Copyright © M.J. Trow, 2022

ISBN 978 1 39909 305 7

The right of M.J. Trow to be identified as the Author of this work has been asserted by him in accordance with the Copyright, Designs and Patents Act 1988.

A CIP catalogue record for this book is available from the British Library.

All rights reserved. No part of this book may be reproduced or transmitted in any form or by any means, electronic or mechanical, including photocopying, recording or by any information storage and retrieval system, without permission from the Publisher in writing.

Printed and bound in England
By CPI (UK) Ltd.

Pen & Sword Books Ltd. incorporates the Imprints of Pen & Sword Archaeology, Atlas, Aviation, Battleground, Discovery, Family History, History, Maritime, Military, Naval, Politics, Railways, Select, Transport, True Crime, Fiction, Frontline Books, Leo Cooper, Praetorian Press, Seaforth Publishing, Wharncliffe and White Owl.

For a complete list of Pen & Sword titles please contact

PEN & SWORD BOOKS LIMITED
47 Church Street, Barnsley, South Yorkshire, S70 2AS, England
E-mail: enquiries@pen-and-sword.co.uk
Website: www.pen-and-sword.co.uk

or

PEN AND SWORD BOOKS
1950 Lawrence Rd, Havertown, PA 19083, USA
E-mail: uspen-and-sword@casematepublishers.com
Website: www.penandswordbooks.com

Contents

Author's Note and Acknowledgements		vi
Foreword	Sefton	viii
Prologue	Eohippus – the Dawn Horse	xii
Chapter 1	A Gallop Through the Ancient World	1
Chapter 2	Bucephalus, the Bull Head	10
Chapter 3	Each, the Horse People	21
Chapter 4	Incitatus, Consul of Rome	32
Chapter 5	A Gallop Through the Dark Ages	48
Chapter 6	Babieca, the Booby	55
Chapter 7	The Wind Horse	70
Chapter 8	White Surrey	79
Chapter 9	Tziminchak, the Thunder and the Lightning	90
Chapter 10	Black Barbarie	97
Chapter 11	A Gallop Through the Age of Reason	109
Chapter 12	Copenhagen v. Marengo	119
Chapter 13	Ronald	139
Chapter 14	Traveller v. Cincinnati	153
Chapter 15	Comanche	165
Chapter 16	A Gallop Through the Empire	178
Chapter 17	Warrior	193
Chapter 18	A Gallop Through Hollywood's Horses	208
Epilogue	'Goodbye, Old Man'	226
Bibliography		232
Index		235

Author's Note and Acknowledgements

Although I have only been on the back of a horse once or twice in my life, I have always held them in the highest regard. My love of all things cavalry began at a very early age and one can't have an interest in military history without also falling in love with the horse. Along with dogs – and, more arguably, cats – horses have been our companions for so long that we may be tempted, perhaps, to take them for granted, seeing them in a field or walking in solemn single file along the side of the road as we drive past. But their link with men is stronger than that – they have borne us to war almost from the outset and our debt to them is great.

The chapters which follow comprise some 'gallops through', which give a general overview of a period of time and help to clarify the more specific chapters based around a single horse, or occasionally a pair from opposing 'camps' such as Napoleon's mount, Marengo, and Wellington's favourite, Copenhagen. The fictionalised sections at the beginning of these chapters take facts from the historical record but are nonetheless told as stories. Because what, after all, is history if it is not a story?

And this is the story of man and his horse – a soldier's mount.

As always, I must thank Heather Williams and her team at Pen and Sword for giving me the opportunity to share my love of the cavalry horse, with especial thanks to my copy editor Gaynor Haliday. My son Taliesin wrote the Gallop Through the Dark Ages for me – we talked so much about the era as I mined his archaeological knowledge that it became clear that it would be simpler if he just wrote it all down! So many thanks go to him, as always. I don't know how many times I have thanked my wife, Carol, for her help, not just in typing the manuscript from my almost indecipherable writing but in all things. I just know that I do thank her, yet again.

M.J. Trow
Vectis
September 2021

Foreword
Sefton

It was a routine day, Tuesday, 20 July 1982, in London. Trooper Michael Pedersen watched the tourists rubber-necking their way to the side of the road. Around him, ahead and in front, sixteen men and horses of the Blues and Royals, resplendent in their polished plumed helmets, back and breastplates, their tall boots gleaming in the sun. It may have been Pedersen's own bias, but he knew that there was no cavalry in the world that still looked like this. Perhaps the Americans watching, smiling, waving, clicking away with their cameras, just saw some limeys in fancy dress – they, whose army had *never* looked like this. But Pedersen knew that he and his comrades, on their way to the changing of the guard from Knightsbridge Barracks, were real soldiers, as at home in khaki as they were in armour, as comfortable in tanks and armoured vehicles as they were in the sheepskins of their saddles.

'Sharkey' tossed his head, champing at the bit. He liked processions, this big, black 17-year-old with the white blaze, but he liked showjumping and point to point even more. His real name was Sefton, named after an Irish peer who had been an officer in the Household Cavalry. Pedersen and the other lads called him 'Sharkey' because he was known to bite. To be fair, that was years ago, and Sefton was an old warhorse now, nearing retirement. Soon, he'd be rolling in the hay while Pedersen was still entertaining the crowds or serving Her Majesty wherever the queen's government decided to send his regiment.

Ten forty a.m. A car parked on the roadside, a blue Austin that Pedersen hadn't noticed, suddenly exploded in a sheet of flame. The impact shattered the windows of other vehicles and rattled house windows yards away. Two troopers died at once, their bodies blasted by the bomb. Two more were ricocheted out of their saddles and would die days later. Seven horses went down and did not get up. Cedric, Epaulette, Falcon, Rochester, Waterford and Zara were all still alive, but they could not stand, rolling in agony, their bodies riddled with the flying nails of the bomb.

Sefton reeled, but Pedersen steadied him, gripping instinctively with his knees, pulling hard back on the curb rein. He couldn't hear anything, certainly not the screaming of the crowd scattering in all directions, nor the dying sounds of the horses. From nowhere, officers and men of the regiment were rushing to the scene. Colonel Andrew Parker Bowles, the commanding officer, issued orders, telling a groom to take his shirt off and to staunch Sefton's wounds. He could do nothing for the two dead men. He consoled the others. Pedersen had dismounted and could only stand there, by his horse's bloody head, staring numbly at the scene of carnage.

No one in London that day, probably no one in Britain, had ever seen dead horses lying in a street. No horse deaths were recorded when the Household Cavalry 'aided the civil power' in 1887, controlling the rioters of the East End, pushing the rabble away from Whitehall and Buckingham Palace. Neither had any animals been killed when in the summer of 1819, the Manchester and Salford Yeomanry had charged a crowd of unarmed men, women and children who had collected at St Peter's Fields in Manchester to listen to the radical tub-thumper Henry Hunt. The last time Britain had seen dead horses was on the killing field of Culloden, Scotland, in April 1746 when 'Butcher' Cumberland had destroyed the ragtag Highland army of Bonnie Prince Charlie. The first cameras on the scene in July 1982 showed the seven dead animals lying in their own blood, their saddles and harness removed. Official press pictures showed blankets over them; the sight would have been too grim for the British public.

Sefton's jugular vein had been severed and he was temporarily blinded in both eyes. He had thirty-four wounds all over his body and was the worst hit of the surviving animals. Veterinary officer Major Noel Carding tended to him immediately and got him into a horsebox, his head steadied by Farrier-Major Brian Smith and three others. In the forge at Knightsbridge Barracks, Carding carried out a 90-minute operation on Sefton as well as checking on the others as they arrived. He was the first cavalry officer to treat a horse for war wounds in half a century.

Sefton underwent a total of eight hours of surgery after his removal to the Royal Army Veterinary Corps at Melton Mowbray, Leicestershire. Thousands of cards were sent by a shocked and furious public who raised an astonishing £620,000 which was used to build a new hospital wing at Melton Mowbray. Later that year, Sefton was back in harness, with Trooper Pedersen on his back, the darling of the crowds who had made him the most famous horse in Britain. He was given the Horse of the Year award in 1982 and retired from the army two years later. He ended his days, as Pedersen always hoped he would, at the Home of Rest for Horses in Speen, Buckinghamshire, along with Echo, a grey of the Metropolitan Police hurt in the same bomb attack. Sefton had to be put down on 9 July 1993 because of lameness caused by the 1982 bomb. He is buried at Melton Mowbray and a statue was unveiled to him in 2013.

As to those members of the IRA responsible, Gilbert McNamee was jailed for twenty-five years for making the bomb. He was released under the Good Friday Agreement in 1998 which freed convicted terrorists in order to create peace in Northern Ireland.

The bomber was John Downey. The case against him collapsed when it was revealed that he had received an 'On-the-Run' letter from Tony Blair's government, assuring him that no prosecutions would take place, either for the Hyde Park Bombing or an earlier atrocity in which he was involved.

Sefton and the horses who died in 1982 stand out in recent history because we do not expect casualties of this kind on British streets. Most right-thinking people regarded the bombing as an act of terrorism, of the sort we have become sadly used to over the last fifty years. The IRA have always regarded their campaign as part of an ongoing war. In that respect, Sefton was a war casualty, like many thousands more referred to in this book. Horses do not understand politics. Whatever animal rights activists believe today, they do not take sides over issues. Religion, power, causes, opinions – these are the creations of men, not horses.

As an anonymous soldier says in the 1995 film *Pursuit of Honor*, '[Horses] fight beside us and die beside us – they deserve the same respect.'

Prologue
Eohippus – the Dawn Horse

Sergeant Milton DeLacey of I Company, United States 7th Cavalry, rode over the battlefield still littered with corpses, butchered in the tall grass, picked at by the scavenging vultures. Coyotes and wolves had crawled as near as they dared, dragging at hands and feet, snarling at each other as they fought over the choicest morsels.

Nobody knew how many men had died fighting the finest light cavalry in the world; the final count had yet to be made. It was 27 June 1876 and the unthinkable had happened; a decorated war hero of the American Civil War, George Armstrong Custer, had died at the head of his companies of the 7th and, to the majority of Americans, civilisation had been stopped in its tracks. Thanks to the proximity of the Union Pacific Railroad and the technology of the electric telegraph, news spread fast, and the journalists of the day leapt, as usual, to simplifications and bias to give their stories more bite. The battle of Little Bighorn, fought along the river of that name, almost immediately became the 'Custer massacre', even though it was a slaughter of the general's own making. And there were no survivors.

Sergeant DeLacey knew different. As he trotted over bodies on that June day, he knew that many men of the 7th had survived, men of Major Reno's companies and Captain Benteen's, who had been under fire but not destroyed as Custer's outfit had been. DeLacey reined in and cocked his carbine. A noise in the ravine to his right had startled his horse and the sergeant was ready for anything. A bay gelding lay in the grass, the McLellan saddle slipped to one side; he was bleeding from seven wounds, waiting patiently to die with calm, quiet eyes.

DeLacey dismounted, still gripping his own horse's reins and stroked the injured mount's muzzle. He hauled down his water canteen and bathed the animal's nostrils with it. The wounds were serious but with care and a little love, he might just pull through. Then, a particular wound caught his eye. It was an old one, a pale scar on the left shoulder and he knew at once that this was Comanche, the horse of DeLacey's commanding officer, the Irishman Captain Myles Keogh. Where the captain was, DeLacey did not know, but in spite of the shock on that desolate, windy ridge, the sergeant was determined to save his horse.

Another piece of journalese followed; Comanche, said the papers for years to come, was the only survivor of Custer's 7th. In fact, he was one of nearly 100 horses the Lakota and Cheyenne had not taken from the field as the spoils of war; and there was a bulldog too.

Nine years earlier and only a few miles from the Custer battlefield, a palaeontologist was tapping at a rock formation in the Wind River Basin, now the home to Shoshone and Arapaho tribes in the United States' seventh largest reservation. Palaeontology was a relatively new science in the 1860s and the territory of the Wind River, which becomes the Big Horn halfway along its length at what the native population calls 'the wedding of the waters', was a focal point of development.

In 1867, while the palaeontologist was looking at fossil remains, other miners were looking for gold. This was Indian Territory, marked as such on contemporary maps, and various treaties between the American government and the tribes tried to ensure that the 'westward march' of white civilisation was controlled and ordered. Gold destroyed all that. Immigrants from all over the world descended on Dakota's Black Hills, as they had on California twenty years earlier. While the prospectors' greedy attack on sacred indigenous sites would lead directly to the destruction of Custer's command and later atrocities, the work of the palaeontologists led to a new discovery – the fossil remains of what came to be known as Eohippus, the dawn horse.

The earliest horse stood about one foot high at the shoulder. One of the foremost American palaeontologists of the 1870s referred to it as being the size of a fox terrier; Henry Fairfield Osborn was a keen fox hunter (never a mainstream American hobby) and the term stuck. Eohippus was, in fact, born to controversy, as part of the 'bone wars' of the 1870s in which palaeontologists, particularly Osborn and Othniel C. Marsh of Yale University, clashed with each other over the origin of various species. Not unnaturally, most of their focus – and that of the press – was given to dinosaur bones, at once spectacular and terrifying. It was Marsh who coined the name triceratops for the extinct animal that is now a symbol of the state of Wyoming, set up in 1890 with Cheyenne as its capital.

This was the era when Charles Darwin's theory of evolution, in his *Origin of Species*, took the world by storm. Thomas Huxley, in Britain, extended Darwin's work to include humans, which Darwin himself was loath to do and, in conjunction with Marsh, Huxley delivered a memorable lecture on the origins of the horse in New York in the year of the Big Horn battle.

Eohippus lived in marshy everglades, long before the Great Plains developed in the early Eocene period, some 39–56 million years ago. A more recent descendant was Mesohippus, the middle horse, about 30–40 million years ago. This animal was four foot long, standing two foot tall at the shoulder or 6 hands in modern equine height terminology, originally established by the polymath Leonardo da

Vinci. Mesohippus' skull was longer than that of Eohippus, giving it a more distinctive horse appearance. The skull had a depression in the centre and the teeth were low-crowned with a space where a modern bit would rest. It had toe-like digits, four on the front foot and three on the back, the third digit being longer than the others, with a soft pad under the toe. In time, the largest digits would become the hoofs and the others the heels. Eo- and Mesohippus were hunted animals, with eyes wide on the side of the head. They lived, as most prey animals did, in herds and their coats were probably striped or dappled to provide camouflage against their predators. A full-grown adult weighed about 50lb (110kg). The most complete skeleton of Mesohippus was found in Wyoming in 1931.

One of the great imponderables of the early horse is how and when an animal hunted as food became domesticated. Several areas of the world claim the honour for this, but it was the Eurasian steppes, north of the Black Sea, that probably saw the first horses ridden by man. Professor Jared Diamond, author and lecturer in geography at UCLA, has listed six essential criteria for an animal to become domesticated. It must have a flexible diet; a finicky creature that only eats one thing is an evolutionary disaster – witness the panda. By the time man was taming horses, their diet included grass that was plentiful and cheap. They must grow fast; a long gestation period and years below full strength and speed is of limited use to men who have plans for such animals. They must breed in captivity; again, we have the panda as an example of a creature that regularly fails in this respect. In societies, like the Scythians and the Huns, where wealth was measured in horses, a stallion must be able to cover as many mares as possible. They must have a pleasant disposition; this is a tricky one, because this book is full of difficult horses which could only be ridden by one hero – Alexander the Great's Bucephalus, for instance – but generally speaking, the horse is a docile creature.

The fifth criterion is a reluctance to panic. We will look at the behaviour of warhorses in battle throughout this book; some were skittish – for example, Sultan, the charger of Colonel Clarke of the Scots Greys at Balaclava in 1854 – but animals bred for war were trained to the alarms and excursions of the field and got used to them. Finally, an animal suitable for domestication must accept a modified social hierarchy – that is, they recognise humans as their masters.

The nearest existing animals descended from the earliest domestic horse are the Tarpans, but today's breed has been watered down by cross-breeding and are not *quite* the same animals we see painted on the walls of the Lascaux caves. Those animals almost certainly pre-date domestication and are shown in the same context as deer and wild cattle – they are animals for the hunt. The Tarpan or European wild horse has never actually been tamed and can be vicious if threatened. Anyone who has been attacked by a horse, or seen stallions fighting, will know that they use hoofs and teeth indiscriminately and use their heads to stun their opponent. Tarpans stand at 12 to 14 hands, are usually of dun colouring with darker legs and a dorsal stripe. Their manes stand erect. Their diet is grass and rhubarb roots. The pure Tarpans were wiped out by farmers in the nineteenth century as they were tired of the animals eating their crops.

A similar 'ancient' breed is the Przewalski, discovered and named by Captain Nicolai Przewalski of the Russian army while travelling in Mongolia in 1878. These were the horses ridden by Genghis Khan and his warriors of the golden horde who terrorised Eastern Europe in the fourteenth century.

The earliest artistic depiction of a horse was found in the Vogelherd cave in Germany. It is a carving on a mammoth tusk, 2½in long and was carved about 34,000 years ago. While both Neanderthals and

Cro-Magnons hunted the horse, there is evidence from the Asian steppes in the early fourth millennium BC of the animals being used for milk as well as meat. This implies domestication, with pens and/or stables but the economy of the Steppe people was nomadic, so they may have moved with their herds. If so, their lifestyle was literally horse-led. Precise focusing on a time or culture for horse domestication is impossible, but most experts today opt for the Botai people of northern Kazakhstan about 4000 BC. From skeleton remains involving large numbers of horses, we know that culling was common (perhaps to reduce the need to feed too many horses) as was animal sacrifice. The horse as god is a theme throughout history – we shall refer to it several times throughout this book. The first reins – merely a rope around the animal's jaw – come from graves in the steppes north of the Black Sea. Mouthpieces (bits) of sinew or hemp date from soon after and are arguably kinder than modern metal versions.

As we shall see, the invention of the wheel took horse warfare in a parallel direction, chariots being more common than conventional cavalry in a number of ancient civilisations.

Whenever and wherever the relationship between horse and man developed, it was a marriage made in military heaven. Not until the advent of devastating artillery and long-range guns did the horsed warrior become obsolete and by then, warfare had reached the skies.

Chapter 1

A Gallop Through the Ancient World

'The Assyrian came down like a wolf on the fold,
And his cohorts were gleaming in purple and gold.'

The horse has always held a fascination for man. Some of the earliest legends and stories concern themselves with spirit animals who appear as horses, as in the *Rig Veda*, a Sanskrit collection of poems or hymns from the second millennium BC. The image of a horse pulling a chariot across the sky as a symbol of the sun is an enduring one.

Across all Palaeolithic cave art, the horse is the most frequently depicted, and archaeological evidence exists of rein and bit as early as 3500 BC. The horse was seen as wild and powerful, and was domesticated not long after cattle and sheep, especially in the Steppe where these hardy animals could endure harsh winters, and be used as beasts of burden, as well as a food source. Their military use is not well documented in early prehistory in Europe, but in Africa and the Middle East pictorial and written evidence dates from the second millennium BC.

Byron's *The Destruction of Sennacherib*, quoted above, was published in 1815, the year of Waterloo, and was a nod in the direction of one of the tyrants of the ancient world. Sennacherib was king of Assyria, one of the many warring states jostling for power to the north-west of the Caspian Sea. In controlling and extending his vast empire, he besieged Jerusalem for fifteen months and destroyed Babylon in 689 BC before one of his sons killed him. He also built

Nineveh as his capital; many of the warhorses he depicted in bas relief on the walls are now in the British Museum.

The Assyrians rode bareback and had no stirrups. Charging into battle using reins and gripping the horse's body with their thighs, troops must have had astonishing skills lost to later horsemen. Bits had to be wider so that the rider could steer more effectively, and this probably damaged the horses' mouths over time. Both plain and jointed snaffles were available by about 1400 BC.

Under Assurnasiraph II, most cavalry were archers, wearing no armour. Bas reliefs from Nineveh and Khorsabad show them leading two extra horses, perhaps to provide remounts for animals killed or because Assyrian horses lacked stamina. There is the suggestion that the Assyrians fought in pairs, one rider handling the animals, the other firing the bow, rather as ancient charioteers fought. By the eighth century BC, in the reign of Tiglathpilser III, these archers wore armour, overlapping plates of bronze stitched to a leather tunic, and carried bronze helmets. The implication, although there is no hard evidence for it, is that the horsed archer was now more aggressive, used for frontal assaults rather than hit-and-run raids. Under Assurbanipal (668–626 BC) three types of cavalry emerged: the lightly armed, unarmoured auxiliaries carrying bows and javelins, heavy archers in armour and fully armoured warriors carrying lances, for thrusting rather than throwing, and swords. The most common breed of horse was probably the Akhal-Teke, of the pale colour called 'buckskin' by nineteenth-century Americans.

But the Assyrian horsemen were not the first to make their mark on history. To the north-west of Sennacherib's state was Mitanni territory, the tribal lands of the Aryans who settled in what used to be called the 'cradle of civilisation', the fertile crescent between the rivers Tigris and Euphrates. The Mitanni were keen charioteers, with pairs of drivers competing in their circuses. The chariot was an important technological development in the history of cavalry warfare, rather as the stirrup was centuries later. The first vehicles, used for hunting and fighting, rather than carrying goods, were probably invented by the Hittites. They drove two or three horses and their wheels had spokes to reduce weight.

The Hittites were not a horse-breeding people, unlike the Mitanni, but they recognised the impact of a war chariot, carrying two, perhaps three archers crashing into terrified infantry on the battlefield. The Hittite use of bronze meant that horse harness and chariot fittings were not only long-lasting and serviceable but have survived to become archaeological artefacts. It is from the Hittites that we have the first record of warhorse training, written in cuneiform on clay tablets *c.* 1345 BC on the orders of Kikkuli, horse-master to King Suppililmer. Kikkuli himself was probably Mitanian, which explains his position at court and as a horse expert of some significance. His training regimen began with the horses being led, at the walk, trot, canter and gallop before being introduced to the weight of a rider or being harnessed to a chariot. The task was arduous with three sessions a day over a seven-month period; compare this with the training given by Captain Louis Nolan to horses destined for the Crimea – sixty-four days of training with constant noise of gunshot and drums to acclimatise them. Day Two consisted of one league (3 miles) at the walk and two furlongs (⅛ mile) at the run. The diet for that day was two handfuls of grass, four of barley and one of clover, with grazing allowed all night. This built up, both in terms of exercise and food until the regime ended. It included care of the horse, bathing, swimming, sweating and covering in blankets.

We have no clear idea of what the warhorses of the ancient world looked like. Bas reliefs from Assyria, Mesopotamia and Egypt all show magnificent animals with flaring nostrils, arched necks, 'clean' legs and short backs. Many of them are stallions, stressing the masculinity and 'gung-ho' status of both horse and rider. The Assyrian archers from the palace of Nineveh in the 630s BC (the reign of Assurbanipal) have long braided hair and full beards. Their tunics are long and they fire short, recurved bows, drawn back to the ear like Medieval longbows. Behind each warrior rode a 'back-up', carrying quivers of arrows to provide more ammunition. Judging by the scale, the horses cannot have been more than 14 hands (142cm) high and, like their riders' hair, their tails and manes were braided.

Despite the terrifying impact of chariots, they had their limitations. To make them light enough to be pulled at speed, they could also be flimsy and could flip over, especially on rough ground. Although dates are naturally hazy, it was sometime around 900 BC that we find mounted warriors on the battlefields of the Middle East. Particularly devastating were the Parthians, whose tactic of firing bows one-handed from the saddle entered the English language. The 'Parthian shot' was the description of a warrior turning in his saddle to fire from behind (or under the neck of his horse) out of range of infantry missiles. Since this was often carried out as the Parthians were leaving the field, it transformed into today's 'parting shot'.

The first description we have of a warhorse comes from the Book of Job in the Old Testament, probably written in the sixth century. The translators working for James I of England to produce their version of the Bible in 1611 never let logic get in their way and some of what follows does not make a great deal of sense. That said, the attitude shown to horses as faithful, courageous servants of man stands out and has never gone away:

> Hast thou given the horse strength? Hast thou clothed his neck with thunder? Canst thou make him afraid as a grasshopper? The glory of his nostrils is terrible. He paweth

in the valley and rejoiceth in his strength; he goeth forth to meet the armed men. He mocketh at fears and is not affrighted; neither turneth he back from the sword. The quiver rattleth against him, the glittering spear and the shield. He swalloweth the ground with fierceness and rage; neither believeth he that it is the sound of the trumpet. He saith among the trumpets Ha, ha; and he smelleth the battle afar off, the thunder of the captains and the shouting.

The Revised Version of the Bible (1885) corrected some of the anomalies above but lost some of the 1611 magic too.

The Israelites had a mixed attitude to their horses. The anonymous writer of the psalms of King David says, 'A horse is a vain thing for safety, neither shall he deliver any by his great strength.' Be that as it may, David's son Solomon had 40,000 stalls (teams) of horses for his chariots in 974 BC and a further 12,000 cavalry. 1 Kings 4:28 records that 'Barley also and straw for the horses and swift steeds brought they unto the place where the officers were, every man according to his charge.'

Egypt had little use for the warhorse until they saw it in action. About 1700 BC the Nile basin was invaded by the Hyksos, the 'Shepherd Kings', probably Semites from the north. The Hyksos used chariots, and the Egyptians, in desperation, introduced the vehicle to their own armies. The temple of Rameses II, famous for using his chariots in pursuit of Moses and the children of Israel to the Red Sea, depicts these chariots, lighter and more manoeuvrable than the Assyrian and Persian versions. Once again, we are in the problematic area of artistic style. The horses, usually two per chariot, are noble, with long legs and arched necks. They wear fringed cloths and are harnessed to the chariot by a yoke. In one example, the reins are lashed around the driver's waist, but only the pharaoh rides alone;

in all the other chariots, there are three occupants, presumably two warriors and a driver.

Long after Rameses' time, Cyrus the Great dominated the Middle East. The epitaph on the tomb of his father, Darius I, reads, 'I was a great rider and a great hunter. For me, nothing was impossible.' This is proof that the status of the cavalry – and of the horse – had changed since earlier times. Horse harness had become increasingly elaborate, decorated with gold and silver; and horses were treated as symbols of power and wealth. Cyrus was the founder of the Persian empire, scion of the Achaemenid dynasty and he took Babylon in 539 BC. He worked with the Phoenicians and Israelites, to the extent that the Old Testament refers to him as the Shepherd and Anointed of Jehovah. The Greek historian Herodotus was in awe of the Persian charioteers – 'no man dared face them'. We have no clear idea of the breeding origin of Cyrus' horses, but Xenophon, the Greek military expert (see Chapter 2) believed they were from Bactria (now Afghanistan). The Greeks called them Niseans, after a legendary hero. The poet Oppian wrote of them in AD 211, 'The horses of Nisea are the handsomest, fit only for mighty rulers. They are splendid, running swiftly under the rider, obeying the bridle willingly.'

Xenophon's description of Persian cavalry refers to the elite, Cyrus' personal guard. They wore bronze helmets with horsehair plumes, itself symbolic of the relationship between animal and rider. Each man had a long, ankle-length coat covered in overlapping scales of bronze, effectively making them heavy cavalry. The horses had armour too – a bronze breastplate and a chamfron to protect the face.

One of the most persistent and defiant of Cyrus' enemies were the Scythians, Iranian nomads from the Steppelands between the Danube and the Black Sea. Centuries later, their leader Timur-i-leng (Christopher Marlowe's *Tamburlaine*) was still causing havoc in the West. They joined forces with the Babylonians and Medeans and when harassed by Darius I fell back, using a scorched earth policy such as Vlad Tepes, the Impaler, would use against the Turks in the sixteenth century, and the Russians against Napoleon in 1812. They burned crops and poisoned wells with the carcases of animals. Scythian tomb furniture has revealed a plethora of horse equipment and skeletons – one has over 400 horses in it. Intriguingly, these occur in women's graves too. The overlapping scales on warriors' and horses' armour were the speciality of the Scythians, whose craftsmen were renowned for the work. They were also unique in the ancient world at that time in that they wore trousers, probably to protect their legs against constant mounting and dismounting. They were particularly impressive archers using a 30in-long composite bow with horsehair string. Experimental archaeologists have rebuilt these weapons, which have a range of 400 yards (353m). An experienced horse archer could fire ten arrows a minute, so that an avalanche of death hit beleaguered infantry trying to make a stand.

The most physically impressive cavalry of the ancient world were probably the Sassanids. First noticed in the sixth century BC, they probably copied the Medeans in horse breeding and specialised in heavy, large animals, perhaps 15 or even 16 hands, the height of horses in the First World War. The Greek word for these heavily armoured warriors was *cataphracti* (covered all over). Such armour was expensive and became the gear of the nobility, adding to the cavalry's status. A shirt of mail (linked rings of bronze, later iron) covered the rider from neck to foot. Over his head he wore a mail hood with eyeholes, which must have made breathing difficult and talking virtually impossible. The body was further covered with a leather tunic with stitched metal plates. The legs had hinged bands, rather like a woodlouse or armadillo, as did the feet. These warriors carried long, straight swords to give maximum reach from the saddle and a 13ft lance. Horses were armoured too, as is apparent from an archaeological find at Dura Europos. The animal's head, neck and chest are protected by a leather covering stitched with plates, 2½in long and ¼in thick. The stitching was bronze even when iron plates were used. Each horse armour had about 1,300 scales and weighed up to 88lb (40kg). Because of this, the equivalent of fifteenth-century European knights, the *cataphracti* attacked at the trot and were at their most effective battering their way through determined infantry.

The Thracians, from what is today Bulgaria, were famous horsemen – the Greeks and the Romans employed them as mercenaries. Like the Celts of the same period (see Chapter 3) they loved drinking, singing and storytelling. Their battle-songs were called *titanismos* by the Greeks, the song of the Titans, the terrifying giants of Greek mythology. The Thracian horseman had such a high reputation that he became a mythical, god-like figure himself, found in shrines and on artwork throughout the empire.

The Thracian cavalry were renowned javelin throwers and they used bows from the saddle too. Their shields are always shown in contemporary art as being strapped across their backs, presumably to prevent attacks from behind. Their swords were curved. Most Thracian armour was copied by the Greeks, especially bronze greaves to protect a rider's shins.

Thracian horses were 14 hands high at the most. Xenophon reported that horse racing was only ever held downhill to be able to record the maximum speed. Their horses were richly decorated with silver and occasionally gold. Actual examples of saddles have been found from the second century BC in southern Thrace. They are low, simple, of brown leather but with highly decorated saddlecloths.

In the Steppe, many groups in modern-day Russia, Ukraine and Kazakhstan were using horses as symbols of power and military might. The Pazyryk culture *c.* 500–300 BC was very reliant on the horse; a large burial mound gives us immense insight into how important they were. Their horse furniture was highly ornate, making them appear almost as supernatural animals, perhaps with both ritual significance and as a way of intimidating an enemy. Felt fragments which have survived in the permafrost show an armed nobleman astride a galloping horse.

The tomb of Qin Shihuangdi (r. 220–210 BC), the world-famous terracotta army, reputedly contained bronze and terracotta depictions of over 600 horses and 130 chariots.

But there are no named horses in the ancient world. Not until Alexander the Great do we hear of a horse whose reputation made him as feared as his master.

Chapter 2

Bucephalus, the Bull Head

Philip of Macedon had got it wrong – and he could have kicked himself. He, the great warrior king to whom all the Peloponnese knelt in obeisance, he whose word was law and whose single, staring eye brought terror to his enemies; he had bought a dud horse! And not only that, the animal had cost him 13 talents, a small fortune for a beast that could not have many years left in him.

Philip had checked the horse's mouth, his teeth, his withers, his fetlocks, the arch of his neck and the carriage of his tail. What he had not done was to ride him – somehow, he had never had the time. But he had watched men who had tried. He was watching one now, one of his horse-masters, perhaps the best of them. The animal stood patiently as the new rider took hold of his mane and, with soothing words, leapt on to his back. The horse reared and bucked, whinnying and shaking his head. The horse-master held on but not for long and ended up as all the others had, in the dust, with sniggers from the court.

Philip snarled and got up from his dais. He crossed the parade ground, sword in hand and pointed it at the animal, still again now and watching him carefully. He saw the bull-shaped mark on his forehead and let the sword fall to his side. 'Well, Bull-head,' he muttered, 'you're no use to me. Take him away.'

The horse-master had brushed himself down and gripped the animal's mane.

'What an excellent horse to lose,' a voice called from the dais. Philip turned. He would know that voice anywhere – it was Alexander, his son. 'For lack of skill and boldness in managing him.'

Philip straightened. The boy was 12, barely finished shitting yellow. And *nobody* challenged Philip of Macedon.

'You can do better?' the king sneered.

Alexander crossed to the Bull-head and stroked his muzzle. The animal snorted and moved away. Alexander followed him, cradling his head and turning him to the east. The horse snickered, then stood still. The boy took a firm hold of the mane and leapt astride him. The horse whinnied, but he did not shy. Alexander gently nudged the belly with his heels and the animal broke into a trot, then a canter, finally a gallop, whirling in the sands of the parade ground.

The prince of Macedon pulled his mount to a halt and slid from his back. The court was not jeering now; they all stood there, behind their king, open-mouthed.

'How …?' Philip was lost for words.

'The sun.' Alexander pointed to the sky. 'The Bull-head is afraid of his shadow on the ground. Point him in the right direction and he's as gentle as a lamb. Why has no one noticed it before?'

Tears of joy trickled from Philip's eyes, the good one and the bad. He hugged the boy, kissing his golden locks. 'My son, my son,' he said. 'Look for a kingdom equal to and worthy of yourself. Macedonia is too small for you.'

There has never been a general like Alexander the Great. Dead at 32, he had conquered an empire larger than the world had ever seen and it is difficult to see beyond the superlatives. His territory extended from his native Macedonia to the north of Greece, across the Hellespont into Phrygia, Cappadocia, Media and Parthia to the banks of the Indus, an astonishing 3,500 miles, west to east. From north to south, it extended from the shores of the Black Sea to Egypt (2,000 miles). Such a vast area, populated by myriad tribes and peoples, was impossible to govern and almost immediately after

Alexander's death, his empire fell apart, his generals squabbling among themselves over territory. Just how such a young man, with relatively little experience, could achieve such astonishing military success is still debated today, but undoubtedly the Macedonian had that grasp of strategy and tactics which is instinctive in some men; Oliver Cromwell, for example, had no experience of battle before 1642 – he had read just two books on Swedish drill formations.

It is worth remembering, however, that Alexander was a man of his time. He accepted the latest Greek thinking (Aristotle was his tutor) that the world was made up of a single land mass surrounded by Ocean, a mythical river. He might have ventured into India itself, but his troops, away from home for years as they had been, mutinied and he was forced to retrace his steps.

Alexander cut his teeth in the vicious politics of the Greek world of the fourth century BC. His father, the one-eyed Philip of Macedon, was preparing for an invasion of the mighty Persian empire when he was assassinated in 336 BC. His son may have been involved in the coup against his father; he certainly benefited from the man's death because he was his heir. Only 18 when he fought his first battle (Chaeronea in 338 BC) where he commanded the left wing, he was an experienced soldier by the age of 20 and launched his astonishing campaign.

According to legend, Alexander had already acquired his famous horse, Bucephalus, the bull-headed. He is the first horse whose name we know, but, as usual, all the details about him come from later writers, Greek and Roman sources who cannot be relied upon in terms of accuracy. Bucephalus is described as a bay, chestnut, black, even skewbald. Nobody claims that he was white, however – the colour of a hero's horse over at least two thousand years. The name, ox- or bull-head is said to have derived from the star and blaze on his forehead, which suggested the shape. When Marco Polo travelled east in search of China in the thirteenth century, he met people in Badakhastan (today's Afghanistan) who claimed that their local breed all had that particular mark. Alexander's reputation survived for centuries. One of Bulgaria's leaders in the fifteenth century was Georg Castrioti, but the Turks, impressed by his generalship, called him Skanderbeg (Lord Alexander) in honour of the Macedonian.

Bucephalus was probably one of a breed of Thessalonian horses that had been used for years by the Greeks. They were described by Aristophanes, the fifth-century BC Greek playwright and are shown on the Parthenon frieze and the head, by Phidias, is now in the British Museum. All Phidias' horses are stallions, which may be allegorical, pointing to the strength and procreative ability of warhorses. In reality, stallions were unpredictable on campaign, tending to be too interested in mares to be relied upon by their riders. The necks are arched and the mane stands upright (probably stiffened with lime). From the size of the riders and their feet dangling below the animals' bellies, they were no more than 14 hands high. Interestingly, the Parthenon horses have no harness; the riders are unarmed and are holding on to the animals' manes and gripping their bodies with their thighs. In reality, both Macedonian and Greek horses were adorned with the skins of leopards and lions, the former still in use in the British cavalry until the 1890s.

But it was Philip, not Alexander, who may have instituted the cavalry charge, a massed body of horsemen moving as one rather than the individual exploits of horsemen who fought single, isolated duels with other mounted opponents. It was Alexander, however, who learned valuable lessons from Xenophon, the fourth century BC horseman whose writings are amongst the most important to survive from ancient Greece. His two books on horsemanship, *Hippike* and *Hippicarcus* (the latter referring to the duties of a cavalry commander) were written about 360 BC and are the first known relating to the horse in war. Xenophon was promoted from an ordinary soldier to general during his career, but he does not tell us about his own life. The ideal cavalry horse, he says, should be 'sound-footed, gentle, sufficiently fleet, ready and able to undergo fatigue and, first and foremost, be obedient.' Difficult mounts (like Bucephalus before Alexander got to him) were playing 'the part of a very traitor'. Xenophon insisted that all horses be muzzled, perhaps because he had seen too many unruly animals biting each other in the cavalry lines. He does not mention training schools for animals, although we know soldiers were put through their paces, like those of any well-trained armies. Above all, he taught gentleness:

> Most people think that this [cruelty] is the way to make him look fine; but they only produce an exactly contrary effect – they blind their horses by jerking their mouth up instead of allowing them to see where they are going and, by spurring and whipping, frighten them into confusion so that they run headlong into danger ... But if you teach your horse to go with a light hand on the bit, yet to carry his neck high and bend his head, you will make him do exactly what he himself delights in ...

It is no accident that Alexander's conquests owed a great deal to the warhorse. In all four of his victories over Darius I's Persians, the Greek cavalry were paramount. Although we should be wary of casualty figures from as long ago as this, Darius lost a reputed 110,000 men at Issus in 335 BC compared with 450 from Alexander's ranks. How many of the conqueror's horses reached Xenophon's perfection, we cannot know. A warhorse's hoofs should ring like cymbals, the bones should be good, with a broad chest, powerful shoulders and the neck should be arched. The head must be small and narrow, with wide nostrils and prominent eyes. The haunches should be wide and the ears small.

The odd thing about Xenophon is that he was writing at a time *before* Philip reorganised his army and before Alexander put it through its paces. Like the later Roman fighting machine, the strength of the Greek army lay in its infantry, especially the phalanx formation of the hoplites. Before the first wars against the Persians (500–479 BC), cavalry units existed only in Thessaly and Boeotia (Thebes), both areas known for horse breeding. There were an estimated 1,200 horsemen in Athens by 431 BC, each man owning his own animal and coming from wealthy backgrounds that gave cavalrymen a sense of superiority that has never gone away. The fighting unit, equivalent to a troop in the last two centuries, was the *phyla*, commanded by a *phylarchos*. Five *philoi* made up a *hipparchoi* (the rough equivalent of a squadron) and there were two of these per cavalry group.

The Spartans, probably the most formidable of ancient Greek warriors, had only 400 cavalrymen, emphasising the importance and superiority of their infantry. Thessalonians formed a square for battle, like British infantry in Wellington's day. The fighting unit was the *ile*, 64 men drawn up in 15 files and 4 ranks. At its head was the tetrarch, with three men behind him, five behind that, seven in the third line and so on, in effect forming a mounted triangular phalanx. This formation could swing right or left, adding to the cavalry's flexibility, and each rider was spaced to allow others to fill in the gap for maximum collision force. The tetrarchia of forty-nine men were arranged liked a saw's teeth so that no single horse and man was in the way of another. For a frontal attack, the Thessalonians charged in column and otherwise formed a rhombus (a mounted variant of the phalanx) to attack the enemy's rear or flank. Such cavalry was divided into heavy and light, but the difference, based on weapons and equipment, must have been minimal. A Greek horseman, from any of the semi-independent city states, wore a bronze helmet that covered most of the face and could be tilted backwards when not in use, carried a round or oval shield and used a sword and javelin, ten to thirteen feet long.

Under Philip, the reorganised cavalry were formed by the nobility (*heteroi* or companions) and wore breastplates, again of bronze and often shaped like chest muscles. The king expected men to serve as a matter of routine, rather as the Medieval feudal system raised its knights. Each 'country' provided an *ile* of 200 men. Interestingly, Philip found his light cavalry, responsible for scouting and foraging, in Thrace (today's Bulgaria) with its mystical reference of 'the horseman' often found at Thracian religious shrines. At Chaeronea, Alexander's first battle, the young prince led perhaps 1,800 Macedonian heavy cavalry against the Theban phalanx, driving it from the field.

Alexander's own campaign against Persia began with 30,000 infantry and 5,000 cavalry. Of the horsemen, perhaps 1,000 were Thracians. At Granicus (334 BC) and Gaugemala (331 BC) the cavalry were decisive. He organised his *heteroi* into a personal bodyguard, not unlike the Praetorians of the Romans, made up of veterans or men with clear skills in combat. In his field manoeuvres, Alexander set a pattern that would be followed for centuries. His Thracians threatened and nibbled at the enemy, mounting lightning raids on camps and settlements and retreating into nearby hills before there could be any retaliation. On the field, they protected the infantry phalanx's flanks. He even had cavalrymen who rode to battle and fought on foot, like the dragoons of many centuries later.

Having crossed the Hellespont somewhere near Canakhale, Alexander paid his respects to the long dead (and mythical) Greek warriors who had fallen at the siege of Troy. Cavalrymen seem to have been noticeably gullible over this (non-existent) war, fictionalised by Homer in the *Iliad*. In 1854, when a young British cavalry officer, Richard Temple Godman of the 5th Dragoon Guards, was sailing to the Crimea, he took time off too, to visit the alleged graves of Achilles and Hector, the fictional opponents in a war that never happened.

Alexander's wars were real enough, however. Sixty miles east of his position south of the Sea of Marmara, the Persian army had taken post on the banks of the shallow river Granicus. They had an estimated 10,000 cavalry and 5,000 Greek mercenaries. Perhaps because of arrogance, the Persians drew up a cavalry screen on the riverbank, but their infantry were too far back to form much support. While his generals urged caution, Alexander drew up the typical Greek (and later, Roman) battle formation of the infantry in the centre and the cavalry on the wings. He himself commanded a mixed cavalry/infantry division on the right and as he advanced, he suddenly changed direction, holding his heavy cavalry back and throwing his light cavalry and infantry forward against the Persian left.

The Persians made the mistake of withdrawing men from their centre to support their struggling flank. Alexander smashed into them at the head of his *heteroi*. According to anecdotal history, Alexander speared the Persian commander, Mithridates, and he went down. Leaders engaging in hand-to-hand combat is the stuff of fiction, from long before Shakespeare onwards, but with both generals leading from the front, it is possible. Alexander's helmet was hacked, slicing off his distinctive white horsehair crest, but he fought on regardless. The cavalry had all crossed the river by now and the infantry followed, drawing up their pointed phalanx.

This was the crucial moment that every general looks for – the split second when morale cracks and panic ensues. First the Persian centre, then the wings, hesitated, reeled back and wheeled their horses away. The Greek mercenaries in the rear of the centre were made of sterner stuff and fought manfully against Alexander's onslaught until, beaten and broken, they threw down their weapons and surrendered. It was the cavalry that had won Granicus.

Gaugemala, three years later, was another example of cavalry, well-handled and trained, winning a battle. This clash, fought in modern Iraq, marked the final defeat of Darius. After it, he either committed suicide or was murdered by his own bodyguard. Alexander's light cavalry Thracians pursued the fleeing king, not giving his rearguard time to reform and make a stand. They had ridden 47 miles in a single day.

Unlike many mythologies, such as the Celts' (see Chapter 3) which was zoomorphic and shamanistic, Greek mythology was really just a series of stories, which became ever more elaborate as time went on. For all the Greek gods had supernatural powers, including shapeshifting, they were essentially humans, with human appearances and human frailties. They squabbled, fought, loved, cheated and lied. The only horse deity in Greek mythology is Pegasus, the winged white stallion who was born of the unlikely union of Poseidon, the sea god and the snake-haired gorgon, Medusa. Associated with water (as Celtic horse-gods were) Pegasus created well-springs wherever he placed his hoofs. In some versions, he carries the thunderbolts used by Zeus, the king of the gods on Mount Olympus. No one could ride him until the hero Bellerophon managed it as one of the mighty deeds attributed to such mythical characters.

The centaurs were rather different. There were dozens of them, with names and back-stories, with the upper body and head of a man and the front, back legs and body of a horse. There are several versions of their mythological origin, but the most common is that they were the result of mating between Centaurus and the mares of Magnesia (Thessaly) which says a lot, perhaps, about the sexual habits of the early Greeks! Celtic mythology too, has such man/horse unions among the high kings of Ireland. The *actual* origin may be from the Minoan culture of Crete. There were no horses on the island in the ancient world and centaurs may actually have been nomadic horsemen that Minoans had seen on visits to mainland Greece. Intriguingly, the Aztecs of the sixteenth century believed that the Spanish conquistadores (see Chapter 9) were half man, half horse, for the same reason.

Unlike the future emperor Caligula's horse, Incitatus, who probably never saw actual action in his life, Bucephalus accompanied his master on campaign. Legend has it that Alexander rode other horses on campaign and on the march and only brought Bucephalus on to the field for the final charge, the *coup de grâce*. The animal was nearly 30 when he was killed in battle against the Indian ruler Potus, somewhere along the Indus. Badly wounded, he limped out of the action, carrying Alexander to safety and the chance to mount a fresh animal. 'Though at the point of death,' wrote Pliny, 'and almost drained of blood, he turned, carried the king from the very midst of the foe and then and there fell down, breathing his last tranquilly now that his master was safe.'

Alexander was said to be distraught. The Greeks were not known for sentimentality towards their horses, but Bucephalus was different. In the horse's honour, Plutarch wrote, Alexander built a city, Boukephala, but it appears to have been lost to the desert. The relationship between horse and master was typified by the legend of the kidnapping of Bucephalus in Hyrcania (the Caspian Sea). Alexander issued a proclamation promising the execution of every man, woman and child in the area unless the horse was returned. Needless to say, he soon turned up unharmed. This tale was recorded by the Roman chronicler Arrian in the second century AD. 'Up to then,' he wrote, 'he had shared Alexander's toils and dangers in plenty, never mounted by any but Alexander himself, since Bucephalus would brook no other rider. In stature he was tall and in spirit courageous.' Arrian claims, however, that the horse was never wounded, but died of exhaustion and old age.

Chapter 3

Each, the Horse People

The general had seen this before, but never on these headlands. Men had told him that the island they called Britannia was the edge of the world and that monsters swarmed in the sea that separated it from Gaul. Well, he had seen none over the last three days, just miles of rolling, grey water and gulls that dipped and soared. Were they cheering the legions on or telling them to turn back?

Now, they were here, despite the storm and the sky as black as night, the beaks of the galleys grinding up the shingle of the beach. Ahead of them, on the dunes with their tufts of sea grass and their pale sands, he could see the enemy. His scouts, men who traded with these people, told him that they were the Cantiaci, a wild and fierce tribe who ate babies for breakfast. Their women peed standing up and their men wore rough leggings of cloth down to their ankles; could anything be more uncivilised?

Around him, the 10th Legion were hoisting their shields into position, clapping on helmets and gripping their spears.

'Orders, Caesar?' a centurion barked. Every extra minute waiting frayed nerves just a little more.

Gaius Julius Caesar was buckling on his helmet. The men called him 'curly' already; better not let them see how fast his curls were receding these days. 'What you always do,' the general said. 'Show them what the sons of Jupiter are made of.'

There was a roar from the ships and men poured out of them like ants, to crash into the roaring surf. Those who lost their balance floundered for a moment, then found their feet and struggled upright. But Caesar was not watching them. He was watching the line of warriors on the hills. They *looked* like the Belgae he had beaten already and yet … A handful of charioteers were roaming the dunes, their buff-coloured ponies cantering over the sand, sending dust in all directions. Standing on their backs or on the chariot's yoke, tall men balanced with ease, jeering and gesturing at the Romans bobbing in the unforgiving sea. Their hair was spiked and stiffened with lime and their bodies, hard and sinewy, swirled with blue circles, the beasts and monsters that Caesar had been told lived in their sea. Behind them, standing, legs apart on the light wooden chariots, the warriors of the Cantiaci stood, oblong shields in one hand, gleaming spears in the other. Their roar was guttural, terrifying and they were not going to give the Romans long.

The officers of the 10th ran up the shingle, slipping on the wet stones in their iron-shod caligae. Behind them went the centurions and the first-javelins, trying to form line on the beach. Caesar fumed. Damn this uneven ground. And damn the Cantiaci for knowing where he would land.

'Aquilifer!' he roared from the prow of the ship and swerved aside as a spear bit deep into the wood.

The eagle-bearer heard the word and scuttled forward, the wind whipping the fur of his bearskin headdress. He took up his position at the head of the first cohort, facing the hell that was snaking its way down the beach. The rattle of the chariot wheels over the stones was terrifying, the horses whinnying and stamping in the thrill of the charge. One chariot swept past the Roman front line, the charioteer leaping from the yoke and slashing the aquilifer's face with his sword. Then, before the man hit the shingle, he was back on the yoke and hurtling away in search of another target.

The Roman wall of overlapping shields could not hold against the speed and impact of those tough little horses and the charioteers who drove them. The optios behind the wobbling lines smashed their men's backs with their gnarled sticks, driving them on and screaming at them to stand their ground.

Caesar's face had not changed. His men were being cut to pieces, but he still had his artillery on board his ships and he would unleash his catapults now. But, Jupiter highest and best, where in the name of all the gods, was his cavalry? The storm had scattered them, who knew where, and he did not have a single horseman with him.

Where were the cavalry?

Gaius Julius Caesar wrote an account of the charioteers of the Cantiaci years after he first saw them. They were ranged along the white cliffs near Deal as his battered galleys ploughed into the shingle. His legions, the VII and X Gemina, were already shaken by their Channel crossing. Ships had been lost, men and equipment thrown overboard. And now, beyond the slippery pebbles and the crashing surf, row after row of lime-haired warriors, some astride their tough little horses, some in chariots, others on foot, waited to greet him.

The noise was unbelievable. There was no order, no steadiness, no control, but that hardly mattered. The terror that Caesar spoke of in *De Bello Gallico* was infectious. Each time a cohort tried to form up, shields to the front, spears probing forward, a dozen chariots would smash into them, destroying their cohesion and scattering bodies. The Roman archers could barely get off their shots before the chariots and the horsemen had wheeled away out of range, taunting the invaders for their uselessness.

When Caesar finally got a toehold a few miles inland into the Kent countryside, a foraging party of the VII found itself annihilated by Celtic horsemen and stumbled back to the relative comfort of the general's main force. Of the Roman cavalry, there was no sign; the Channel storms had scattered them, animals and riders slipping below the breakers.

His tail between his legs, the greatest general in the Roman world sailed for home, contending later that he had never intended an invasion of Britannia; it was a punitive raid because the Cantiaci were helping their Belgic relatives in the war that Caesar was fighting in Gaul. There can be no doubt that the horse was central in this debacle – the Roman cavalry failed to arrive, and the Celtic cavalry terrified the infantry who faced them. If the tribes of Celtic British nearest to Deal – the Regni, the Belgae, the Atrebates and the Catuvellani – had joined the Cantiaci as later defenders of Britain did, there might have been no Roman invasion at all. But that was not then the British way – and the Romans knew it.

'One horse was lithe and swift-leaping,' wrote the anonymous poet of *The Cattle Raid of Cooley*, 'high-arched and powerful, long-bodied and with great hooves. The other flowing-maned and shining, slight and slender in hoof and heel.'

Our problem with our understanding of the Celts and their horses is that neither had a written language. In terms of literature, all we know about the people who faced Caesar and the later Roman invasions, is written by the Romans themselves. In the early first century AD, the squint-eyed philosopher Strabo wrote of the Celts:

> The whole race ... is madly fond of war, high-spirited and quick to battle, but otherwise straightforward and not of evil character. And so when they are stirred up they assemble in their bands for battle, quite openly and without forethought ... To the frankness and high-spiritedness of

their temperament must be added the traits of childish boastfulness and love of decoration.

The Celts were an ingenious, inventive people, famed for their beautiful art of the Halstatt and La Tene periods in the 3,000 years before Christ. Nomadic Indo-European warriors reliant on the horse as a means of transport, they dominated the whole area from Ireland, via France and Germany, to Spain. They were fierce warriors, famed oral storytellers and loved nothing more than horses and the sweet drink the Romans stole from them, mead. According to Caesar, the Gauls in today's France and the Britons were divided into three broad classes: the Druid priests; the *equites* (mounted warriors and landowners); and the plebians. This is probably too simple, and it reflects too much the Roman pattern of society at the end of the Republic. The important point, however, is that the *political* leaders of society were horsemen. And the horse was a symbol of power, status and wealth.

Many examples of chariot graves have been found across Europe, almost certainly of high-born warriors whose military equipment was buried with them. Wheels, axles, bits and horse furniture have survived, even if leather has rotted and horse bones are found in relatively few. This may have been because expensive, well-trained warhorses were too valuable a commodity to consign to death, although, as we shall see, Celtic mythology is full of stories of horses dying nobly with their masters.

As the Roman Empire grew and the Roman army was exposed to a wide range of cultures and religious beliefs, soldiers gladly took foreign gods as their own. The Persian deity Mithras is a classic example but the only British goddess to be worshipped openly in Rome was Epona, the horse deity.

We know that the Celts were highly superstitious with their worship at sacred groves involving water, the oak, the mistletoe and the various cults kept alive by the Druids, the mysterious priests who had their cultural centre on Mona, the island of Anglesey. Epona was one of many deities and is sometimes shown as a zoomorphic horse or as a woman riding a horse or with a horse beside her. She is sometimes shown riding side-saddle, a fashion which would not become the norm in Britain until the fourteenth century. Neither was Epona confined to Britain. The Gauls worshipped her under the names of Anu and Danu and it may be that the Epidii tribe, living in what is today Kintyre, held her as their principal deity. It may also be that Epona was Gallic originally and that she was venerated as far east as Thrace (Bulgaria) where, as we have seen, a legendary figure, a horseman, has been found in a number of shrines.

The Gallic word for horse is '*each*' and in various Scottish lochs, the Each Uisge is a siren, half woman, half horse who lures oarsmen and fishermen to their watery deaths. In the Celtic religious year, a horse-goddess opens the gates of life at Beltane in the spring, the traditional mating season, and closes the gates at Samhain in the autumn to signify the coming of winter. A vague folk-memory of the importance of the horse from this pre-Christian period exists today in the hobby horses of the traditional mummers and Morris men. Padstow in Cornwall and Minehead in Somerset both feature these animals as part of the May Day (Beltane) celebrations. Likewise, the Hodden Horse of Kent, the Wild Horses of Cheshire and Shropshire and the Mari Llwyd of Wales all celebrate the end of summer. For centuries, blacksmiths in rural areas held a special place in their communities. At traditional Celtic festivals, the smiths drank first as a token of respect; in Ireland, Goibhniu the smith was worshipped as a god himself.

Celtic horses are everywhere in art. The archaeologist Barry Cunliffe wrote of late prehistoric Europe that the place was 'swamped with bronze horse-gear' much of which is on display at museums in every European country. The hillside chalk carvings at Uffington and Calne are only the most famous and visible of dozens that once bore witness to the horse's importance to the Celts. On one of the best-known artefacts, the Gundestrup cauldron from Rævemose in Denmark, mounted warriors ride around the cup's rim. The animals' tails appear to be heavily braided and they have elaborately decorated harness. In keeping with all the cavalry in the ancient world, there are no stirrups and the jury is still out on the size of these animals. Their riders' feet almost touch the ground – is this because they were small, perhaps 12 hands high? Or is it to accentuate the terrifying hugeness of the warriors themselves?

The horse as a religious symbol is reflected today in the horseshoe. With its 'open' end uppermost, it spelt protection against the devil. Similarly, horse-brasses were worn by the animals to shield them and their owner from the evil eye. Both ideas were probably Celtic to begin with. Giraldus Cambrensis, the Welsh chronicler-monk of the twelfth century records that in Ireland, the kings of Ulster 'mated' with a white mare (which would almost certainly be a criminal offence today). In this inauguration ceremony, the king effectively became a stallion, linking his forthcoming reign with the cycle of fecundity, birth and plenty. The mare was then killed and cooked and the king sat in the cooling stew, eating the flesh and drinking the broth.

The idea of the stallion as procreator is echoed in Irish mythology. Fergus, for example, the son of Ro-ech (the great horse) was extraordinarily well-endowed and it took seven women to satisfy him. Etain Echraide was the horse-riding goddess wife of the god

Midir and the king Eochaid Airem. Echdae was a horse god and Láir Derg was the red mare.

There can hardly be a more colourful character in Irish folklore than Cu Chulainn, the Hound of Ulster. Endowed, as were all Celtic heroes, with supernatural powers, he drank ale and mead by the barrelful, boiled water with his natural body heat, hacked rock formations with axes and swords. And of course, the heroes killed their enemies by the hundred. At Cu's birth, he was given two horses, equally possessed with powers: the grey of Macha and the black of Saingliu. Macha was the crow-goddess, associated with the dead of the battlefield. No doubt Celtic warriors had seen rooks and ravens descend on corpses, pecking at the eyes and soft flesh and they imparted a supernatural significance to them. The link between birds, women and horses is widespread in Celtic mythology and survived into the Viking era in the form of the Valkyrie who swooped over battlefields to take heroes' souls to Valhalla. Although she is far removed from the concept of the horse-goddess, Mother Goose of nursery rhyme fame is the last survivor of this tradition.

The horse is linked, too, with the Celtic cult of the head and several folktales of the Fianna, the most famous of the Irish warbands who were also famous hunters, talk of collecting heads. A number of graves have been found all over Europe that contain a single human head. Traditionally, warriors decapitated their enemies (as the nineteenth-century North American plains Indians took scalps) and hung them from their saddlebows.

In the Welsh tradition, Epona was Riannon, who rode, according to the *Mabinogion*, a pale horse with its biblical connotations. The *Book of Taliesin* from the twelfth century – but based on tales which are much older – mentions a number of supernatural horses that can carry seven men. Black Moro was among them, along with Cornan (the little horned one, who may be a unicorn) and Heith. They all have legendary speed, outdistancing ordinary animals and take to the water like fish. The seahorse (genus *hippocampus*) is a reminder of the amphibious legends of the Celtic triads.

We see Celtic horses in reality in AD 60 when Boudicca, queen of the Iceni, went on the warpath. Her husband, Prasutagus, was a client king of the Romans whose invasion had finally taken place eighteen years earlier. On his death, however, things fell apart. He either expected Boudicca to assert Icenian independence or she resented Roman interference in her rule. Either way, she fell foul of the Roman authorities. She was beaten and her (probably teenaged) daughters were raped. The queen's retaliation was swift and terrible. The Iceni had a reputation as horse-breeders in their homelands of East Anglia, but the account of their rebellion comes from Dio Cassius who was not only a biased Roman but was not there at the time.

'Buduica, at the head of an army of about 230,000 men, rode in a chariot herself and assigned the others to their several stations.' His version of the queen's speech before her last battle runs to four pages, all of it fiction. The fact was that powerful women appalled and frightened the Romans in equal measure, which is why Cleopatra was known as *monstrum fatale* (deadly monster). Cartimandua, queen of the Brigantes who was a contemporary of Boudicca (and whose name means 'sleek pony') escaped censure because she was a good client queen and did as the Romans told her.

There was nothing wrong with Boudicca's strategy. She hit Camulodunum (today's Colchester) because it was a veterans' settlement, full of soldiers. Next, she flattened Londinium, then a wild, timber-built frontier town along the Thames. Not a little rattled, the authorities sent a legion to stop her. But it was not to be. She wiped out a vexillation of the IX Hispana and swung north-west to destroy Verulamium (St Alban's).

With three towns burning and the best troops in the ancient world given a bloody nose, there had to be a reckoning. Having destroyed the Druid stronghold on Mona, the Roman commander Suetonius Paulinus marched south-east with two legions and met the Iceni at an unknown spot (perhaps Mancetter) which came to be known as Manduessedum, the place of the chariots. If Boudicca's strategy was good, her tactics were not. All Celtic armies fought in the same way. Each warrior, mostly without armour, fought his own battle. There was no structure, no cohesion, just wild enthusiasm, courage and a lot of noise. Raucous battle cries, clashing shields and the harsh croak of the carnifex, the battle-horns, was intended to terrify the Romans. But the Romans had seen it all before. They advanced steadily, shields overlapping, spears bristling, in deadly silence and braced themselves for the shock when Boudicca's line hit them.

No doubt the charioteers performed exactly as Caesar had seen them do thirteen decades earlier, dancing on their poles, hauling their tough little horses left and right and jabbing at the legion's edges. There were no knives on the Iceni chariot wheels – that was the stuff of battlefield legend from a later generation. How many died in this grimmest of bloodbaths we do not know. Dio Cassius gives ludicrous figures of 80,000 Celtic dead, but this is hyperbole and typical of the Romans who never met an exaggeration they did not like. There can be no doubt, however, that Manduessedum was a mortal blow to the Iceni and their queen. Boudicca probably survived the fight but disappears from the record soon after and her grave remains unknown.

The horses of the Celts were short-legged and tough, standing perhaps 13 hands high, 1.42 metres from the ground to withers. Their saddles were square, with four projections at the corners, to hold the stirrupless rider steady and they were used for hunting and pack work off the battlefield. With our modern breeds and centuries of military development, we would not be very impressed with the horses that charged Roman legions, but in their day, they won the hearts of men with their stubbornness, their strength and their loyalty; how else can we explain the existence of the Celtic horse in so many examples of literature, mythology and faith?

Chapter 4

Incitatus, Consul of Rome

'This way, sir.'

Quintus Sutorius Macro knew the way perfectly well. He frowned at the lackey the emperor had sent to fetch him from the palace gate. Didn't the man realise who he was? He was Prefect of the Praetorian Guard, for Jupiter's sake, not some underling come to grovel at Caligula's feet. All right, the emperor's name was actually Gaius Julius Caesar Germanicus, but when he was a boy, his father dressed him up as a little soldier to review the troops and he wore a tiny pair of caligae. So that became his name – Caligula. Little Boots. And the emperor hated it. And Macro knew he hated it. And that gave Macro a chance to snigger every day.

The lackey showed Macro into the atrium. The emperor's orders had been explicit; no need for a guard of honour tonight. Do not bother with full dress – armour is *so* tiresome, isn't? Come in a toga – no purple, of course; that went without saying.

'Wine, excellency?' Another lackey had appeared, silver and gilt ewer in hand.

Macro nodded and accepted the goblet. 'Who else is coming?' he asked the steward.

'Just one more guest, sir,' the man said. 'He's here already.' And he bowed and left.

Typical, Macro thought. He had *made* the emperor. If it were not for him and his Praetorians, the weird little nephew of the even weirder Tiberius would remain forever on the fringes of Roman politics, an embarrassment to all he met. The little boy of the little boots had grown into an odd-looking man, with virtually no hair on his head but plenty all over his shoulders, back and chest. He looked like a goat; to the extent that the g-word could not be spoken in the palace. And now, he had only invited one more guest. It *had* to be Gemellus, the boring little shit. If there was anybody more ghastly than Caligula, it was his annoying nephew.

'Quintus Sutorius Macro,' yet another steward called. 'Dinner is served.'

Macro sighed and downed his drink. Say what you like about the emperor – and Macro did – he ate well. Pheasant, peacock, stuffed dormouse; you name it, Caligula provided it. The commander of the Praetorian guard marched, as was his wont, into the Presence. There was no one there. At least, that was not *quite* true. At the far end of the long table, littered with gold plates and goblets, a white stallion stood tethered to the wall. He wore a gold halter and a purple cloth over his back. He looked up at Macro's entrance and went back to chomping his oats.

'Ah, Macro.' Caligula swept into the room, a girl on each arm.

'Sire.' Macro bowed. At least there would be some female company.

'Oh, you know my horse, Incitatus?' Caligula asked.

'Er ...'

'I'm thinking of giving him a seat in the Senate. He's at least as intelligent as most of the jackasses there, don't you think?'

Macro laughed. But Caligula didn't. Neither did his women. Neither did the horse.

The emperor stood apart from his companions. Then he crossed the tiled floor to Macro, looking him squarely in the face. 'You know that job I offered you the other day – Prefect of Egypt?'

'Indeed, sire. I am packed and ready to go.'

'Yes, well, forget that. I've changed my mind. This instead.' Caligula whipped his dagger from his toga and, reversing it in his hand, held the pommel out to his Guard commander. 'What's fine, what's Roman, Macro. You understand?'

The horse shifted in his gilded stall, lifting his head to see Macro's reaction.

'You ... you want me to kill myself?' the Prefect asked.

'Yes,' Caligula beamed. 'That's it exactly.'

And he did.

We know that he was a white stallion, the kind of horse that became equated with heroism down the centuries, culminating perhaps in the 1950s cowboy legends, Hopalong Cassidy, who rode Topper, and the Lone Ranger, in the saddle of Silver. After that, almost everything we know about Incitatus, like his master the emperor Caligula, is hyperbole and myth. The name means swift, in the sense of spirited – the English word incite is linked with it and we can be sure that Incitatus was one of several horses that the emperor owned.

Caligula grew up to be a warped, voyeuristic sadist – not surprising, since he was effectively brought up by his equally warped uncle, Tiberius. In the early years of the emperors, the man who wore the purple had to be careful. The first wannabe ruler of Rome, Julius Caesar, had been stabbed to death for his ambition and the man who finally succeeded him as the first genuine emperor, Augustus, was a master of public relations. He claimed consistently, as did his immediate successors, that he was the people's choice. His legions and his buildings carried the letters SPQR – *Senatus Populesque Romani* (the senate and the people of Rome) as though he was an elected president rather than a despot.

Caligula had no such finesse. He had nothing but contempt for the complicated tiers of government that the republic had created – and thereby hangs the story of his horse. We have two major sources for Caligula's reign: Gaius Tranquillus Suetonius, a lawyer and biographer who put stylus to parchment on the emperor nearly a century after his death; and Lucinus Cassius Dio, the son of a senator, writing in Greek even later than Suetonius. In other words, as is so often the case of ancient world biographies, there is no contemporary account of Caligula's reign, merely a collection of stories, many of them apocryphal, almost all of them hostile, and we have to pick the bones out of them as best we can.

Both writers deal with Incitatus. Cassius Dio says that the animal had his own servants and was fed with oats mixed with flakes of gold. Alarm bells should sound at once. All emperors' horses had their own grooms, so there is nothing unusual about that. If literal, the use of gold as part of a horse's diet would do it no good at all and we can assume that Incitatus merely had the best food available. Dio goes further, however, claiming that Caligula made his pet a priest. The emperor was pontifex maximus, the highest priest in the Roman state, a sort of pope before the papacy was invented. As such, he appointed the rest of the priesthood.

The choice of a horse fitted exactly with the story related by Suetonius, that Caligula planned to make Incitatus first a senator, then a consul. His stable, Suetonius claimed, was made of expensive marble, his manger was ivory, his blankets were imperial purple (only the emperor was allowed to use this colour) and his bridle was studded with precious stones. Dignitaries were invited to dine with Incitatus, but probably his after-dinner conversation was rather limited!

Although he began his reign in the summer of 37 with care and circumspection, Caligula quickly became very disturbed. Some said a sudden illness in October caused this and most horrified contemporaries believed him to be insane. While it was standard to identify dead emperors as gods, Caligula had himself declared immortal. He was Jupiter Latiaris and a body of priests, perhaps with Incitatus among them, was responsible for ceremonies officially worshipping him. As the emperor's behaviour became more wild and unpredictable his list of crimes grew. In May 38, he ordered Quintus Macro, Prefect of the Praetorian Guard who had backed the emperor's accession, to kill himself. Likewise, he had a potential rival, Gemellus, Tiberius' grandson, executed.

There were rumours, which may be based on fact, that Caligula had sex with all three of his sisters. When his favourite, Julia Drusilla, died, he ordered public mourning throughout the empire and had her deified. Such an honour had never been granted to a woman before and Rome read this as another example of the emperor's insanity. Renowned for his sexual conquests, Caligula thought nothing of seducing wives of politicians and generals, often at formal dinners which he hosted. Being aware of a plot against him in 39 involving his remaining sisters, he exiled them both.

Caligula spent money like water, including the lavish arrangements for Incitatus, and inevitably, taxation rose. His mood swings grew worse and by 41 he had managed to offend just about everybody, his name synonymous with cruelty, lust and voyeurism. If Caligula was not actually insane, he was certainly a sadistic narcissist of epic proportions.

One person he should never have offended was Chaerea, a Praetorian tribune who had a high voice and was publicly ridiculed by Caligula for effeminacy. *Quis custodes ipsos custodebat?* the phrase ran in Rome. Who guards the guards? The Praetorians were an elite fighting force, the emperor's bodyguard, and they made and broke emperors. Having backed Caligula in 37, they now turned against him, ambushing the emperor, his fourth wife and his daughter in an alleyway behind the palace. 'The miscreant emperor,' wrote historian Matthew Dennison in *The Twelve Caesars*, 'would die a thousand deaths, victim of a frenzy of killing; his jaw split, his groin ripped by swords, his body bludgeoned, battered and butchered.' Nobody wept for Caligula. We do not know whether Incitatus missed him. We do not know whether Incitatus was still alive.

Twenty years after Julius Caesar's invasion of Britain, the Roman poet Virgil described a warhorse:

> His neck is carried erect; his head is small; his belly short; his back broad. Brawny muscles swell upon his noble chest. A bright bay or a good grey is the best colour; the worst is white or dun. If from afar the clash of arms be heard, he knows not how to stand still; his ears prick up, his limbs quiver; and, snorting, he rolls the collected fire under his nostrils. And his mane is thick and reposes tossed back on his right shoulder. A double spine runs along his back to his loins.

The Roman army was the most successful in the ancient world, but its strength lay in its legions, and they were composed of infantry. Time and again, in battles along the edges of Rome's ever-extending territories, 'barbarian' tribes roared into battle, destroying themselves against the great, silent wall of the legionary shields, going down before the short stabbing sword, the gladius, and being trampled by, if the legion were at full strength, 12,000 feet.

The cavalry, by contrast, were the weak link, although recent work by experimental archaeologists has done something to balance this viewpoint. The horsed units were called *alae* and they fought on the wings of most battle formations. Because the Romans were not natural horsemen, many of the auxiliary cavalrymen were recruited from the very tribes the Romans had defeated. For instance, in Colchester Museum (originally a Roman temple) is the tombstone of Longinus Stapeze, a member of ala I Thracum, a Thracian regiment obviously serving in Britannia. The inscription tells us that he served for fifteen years and died at the age of 40, something of a 'grand old man' for the cavalry of those days. The tomb itself was half destroyed in Boudicca's attack on the town, so we can assume that Longinus was among those who invaded in 43 under Aulus Plautius. He carries what appears to be a circular shield but whatever weapon he was once carrying in his right hand (sword or spear) has gone. The horse is triumphantly trampling over a beaten enemy, lying curled under its hoofs. The harness is elaborate with metal bosses and leather fringework and Longinus was probably an officer of some standing. In common with most of the depictions of Roman cavalrymen, he wears a tunic of metal plates stitched together to form overlapping scales. This gave flexibility of movement as well as protection in action.

As was to be expected in an empire that survived for 500 years, battle formations, structures and tactics developed over time. According to the Greek historian Polybius, the army of the consulate had changed little since the third century BC. Each legion was accompanied on campaign by two *alae* of auxiliary (foreign) cavalry and two of Roman cavalry. As time went on, the *alae* increased in number, but originally there were 300 horsemen compared with 4,200 infantry. The horsemen themselves were usually from the ranks of the elite, the patrician landowners who dominated every aspect of Roman life, be it politics, religion, law, business or the army. Because Roman civilisation remained the ideal for centuries after the empire collapsed, this 'them and us' attitude between the mounted warriors and the man on foot survived for as long as horses were used in war. The cavalry were called *eqites* and the phrase 'equestrian family' was used throughout Europe as a token of superiority and respect.

Each unit was divided into ten troops (*turmae*) commanded by three decurions ('leaders of ten').

Should a horse be killed on campaign, its owner was entitled to compensation to replace it. The *alae* were organised rather differently, divided, like the infantry, into cohorts, commanded by prefects (*prafecti sociorum*) and the Latin says it all; the *socii* were the plebs, the common people, as opposed to the high-born patricians. Their status never reached that of genuinely Roman cavalry and they often fought in mixed units, deploying on foot and carrying out reconnaissance, as dragoons and light cavalry did in later centuries. In the early years, all Roman soldiers were amateurs, a spirit which lasted for centuries at the top of the army; generals were not men of vast military experience, but wealthy patricians who followed the *cursus honorum*, the honourable path of service first in the consulate, then the empire. It is from this that the British purchase system of Queen Anne's reign developed – officers of cavalry and infantry had to buy their rank because they could afford it.

The reforms of Caius Marius, a consul in 107 BC, created the professional legion we can recognise from dozens of Hollywood films. The eagle became the symbol carried by the *aquilifer* (the eagle-bearer) and each unit was given a name, a number and a permanence that had not really existed earlier. The *alae* of this new army had the names of places in which they were raised or had served. For example, Cohors I Britannica, raised under either Claudius or Domitia or Cohors II Augusta Nervia Pecensis under Trajan. Increasingly, the recruiting grounds were Greece and Sicily, Spain and North Africa. Under Augustus, the first emperor, the *alae* became a reliable and highly professional force. On the march, a full unit of these men, a quingeniary (500 men), was impressive, sixteen turnae of thirty men each, led by the prefect and his 'staff', a standard bearer and trumpeter. Not much changed in many ways in eighteen centuries – the Light Brigade drawn up at Balaclava in October 1854 would have looked very similar, even down to the structure of three men riding abreast. General Custer's cavalry on the Big Horn would have recognised this formation too.

In the principate, beginning with Augustus, the equestrians were elevated to key posts in all areas of administration. Commanders of *alae*, often miles from Rome and living in potentially difficult areas like Britannia, had considerable initiative and literally had to be reined in by the emperor from time to time. Such men had considerable glamour too, associated in later times and other countries with the Life Guards in Britain, the Reserve Cavalry under Joachim Murat in France, and the Guard Cossacks of the last Tsars in Russia. The most elite of this elite were the commanders of milliary *alae*; there were never more than a dozen of them at any one time.

The tomb of Vonatrix, son of Daco, from Bonn in the first century AD shows a *spatha*, a long cavalry sword, suspended from a belt at his right hip. Some *spathae* of thirty-six inches long have been found, as long as a Medieval broadsword or a sixteenth-century rapier. Such a weapon was essential in fighting from the saddle. Next time you watch *Spartacus*, notice how awkward Kirk Douglas looks in battle against the Romans; he is trying to use the shorter *gladius* and he can barely reach anybody on the ground!

The prestige of Roman cavalry was preserved by the higher pay going to the *alae*. For instance, under Augustus, a legionary received 225 *denarii* a year; the *ala* equivalent was 262.5, almost a fifth more. As with all armies, these are paper figures. In reality, wages were often late and sometimes not paid at all. Mutinies in the principate's army, where they happened at all, were usually about pay. We would find discipline and punishments, such as decimation (the killing of every tenth man in a unit) unbearable, but most soldiers put up with all that, as long as the gold kept coming. Even so, there were a number of deductibles, so that government coffers were not *too* depleted. Food had to be paid for as did hay for the horses (the legionaries, of course, did not have this problem). One interesting deduction was for *Saturnalia* of the camp. The *Saturnalia* was the midwinter solstice (Christians stole it later as the birth of Christ) and it was associated with drunkenness and parties to welcome the new year. Presumably, it was the rough equivalent of mess bills today.

Cavalrymen, like the infantry, served for twenty-five years, assuming they lived that long, and from Augustus's reign for the next two centuries, were forbidden to marry. Cohabitation was different, although we have little idea where a soldier's common law wife actually lived and children were born of these relationships, especially on the frontiers where fraternisation with the locals became more common as time went on and the Roman occupation was accepted as a fact of life. A famous letter from one sister to another comes from Vindolanda, one of the largest forts along Hadrian's Wall, inviting Sulpicia Lepidina, wife of the commander, to a birthday party.

By the late first century, a typical cavalryman would wear a shirt of mail (links of iron) which was more flexible than the earlier scale armour. His helmet was a variation of the infantry Gallic type, but the cheek plates were higher, and the ears covered. While the infantry could expect an attack from the front, in the whirling melee of a cavalry fight, a sword blade/spear could come from all angles. Officers often had very ornate helmets, decorated with fake hair made of copper or even the real thing. For parades, face masks were worn, the forerunner of visors of the late Middle Ages. Most cavalrymen were armed with spears, used for thrusting over-arm, and a javelin for throwing. They were the exact equivalent of the *pilum* and *hasta* carried by the infantry.

The tombstone of another Longinus, this time of the Ala Sulpicia civium Romanorum, shows horse harness particularly clearly. Longinus is on foot, carrying his spears and walking behind his horse. It is difficult to be accurate about the size of horses from Roman grave furniture, because we have no idea if the man depicted was particularly tall or short. Longinus' horse's ears are on the same level as his rider's head. The four-pronged saddle, with metal projections covered in leather, which gave stability to the rider, is very obvious. The horse cloth is hung with long, dangling tassels and the breast strap has metal discs along its length. The reins appear to be single, because a second rein is probably a training rope held by Longinus himself.

Armour for horses was rare, certainly in the West. Cataphracts, the heavily armoured cavalry of the east, were not raised until the second century under Hadrian. Ala I Gallorum et Pannoniorum Cataphracta left at least one example – a full-size suit of scale armour that covered the horse from head to hoof.

Some time around AD 100, conventional horseshoes came into use, nailed to hoofs. The Greeks did not shoe horses at all (any more than the later Mongols did) and the cavalry of the Roman Republic used 'hipposandals', iron boots tied to the legs as caligae were bound around legionaries' legs. They chafed and let in stones and dirt and cannot have been ideal, especially on long marches.

Unique to Roman cavalry were the sports or games they organised. 'Bread and circuses' appealed to the Romans, from patricians to plebs, and a fortune was spent on watching chariot races in various circuses across the empire. Betting was furious and fights often broke out over various chariot teams rather as fans occasionally resort to fisticuffs on the football terraces today. Parade armour was worn, by horse and man, like the chamfron (face plate) worn by an officer's mount at Vindolanda. The only written account of such games comes from Flavius Arrianus, the second century biographer of Alexander the Great. A warrior himself, driving back an attack by the barbarian

Alans, he describes the long, flowing yellow horsehair plumes of the participants. Eighteen centuries later, the United States cavalry adopted the same colour.

The horsemen went through their paces, attacking each other rather like the Medieval melee, horse against horse and man against man, doing their best to avoid injuries in those 'friendlies'. In what would become pigsticking in India centuries later, cavalrymen threw javelins at a moving target. Fifteen hits was a good result; twenty was exceptional.

High days and holidays like this were important for morale and practice, but most cavalry days were all about routine and boredom, as in any army at any time. One report that has miraculously survived about camp life comes from Cohors I Hispanorum Veterana from *c.* AD 105. There were 119 men under four decurions which, together with the infantry of the legion, added up to 596. As the year went on, some were transferred to Pannonia (not a clear-cut area today, but essentially south and west of the Danube), one was killed by robbers, one deserted. Among the missions men were sent on, some were to get replacement clothing (from Gaul), others to replenish the grain supply. Horses were brought from the River Erar (now unidentifiable), but whether those were trained animals from another *alae* or wild horses to be broken is unclear.

Cavalry were particularly important in patrolling the empire's frontiers. Although the forts and milestones along Hadrian's Wall were close enough to provide cover for each other, probes north into Gododdin country and even further, into Caledonia, were best done by a fast, mobile force. In North Africa, specialist cavalrymen called *dromedarii* rode camels.

By the late fourth century, a number of barbarian tribes were nibbling at the empire's frontiers. Later emperors could not compete with the cruelties of a Caligula or the petulance of a Nero, but neither did they have the gravitas and military ability of Marcus Aurelius,

Trajan or Vespasian. Instead, a succession of inadequates squabbled among themselves and left little mark on the greatest empire in the world. The cavalry that patrolled the frontiers increasingly adopted the tactics and appearance of its enemies. The *spangenhelm*, with its nasal, domed skull and one-piece cheek and neck defences, became the helmet of choice and it was indistinguishable from that worn by the Goths, the Vandals and, the greatest cavalrymen of them all, the Huns.

Ironically, our knowledge of the late Roman Empire is less detailed than, say, for its beginning 400 years earlier. There may have been *comitatenses*, mobile field armies that had no fixed bases but had a strong cavalry arm to be able to get to trouble spots quickly. In the west, still dominated by Rome, were five divisions scattered geographically, a total of seventy-nine units. Britannia was under the control of a Comes (Count) with four vexillations and one full legion. Even this small garrison was depleted from the 490s as troops were pulled back to Italy to defend Rome itself. One new departure was that the cavalry now provided archers, a tactic learned from the enemy, especially the Parthians, who could fire their recurved bows one-handed with great accuracy from the saddle. Not until the American Plains tribesmen of the nineteenth century would the west see such skill.

There is a confusing number of names for the Roman cavalry of the fourth century. Several units were *cataphracti*, heavily armoured. Others were *clibanarii* (from the Latin for a bread oven, which might hint at the discomfort and claustrophobia involved in wearing armour). Successors of the *alae* were the *scutarii* (shieldmen), *promoti* (the preferred) or *stablesiari* (which perhaps speaks for itself, but hints at reliability and steadfastness). It may be that we have the beginnings of heavy and light cavalry which would be applied to all European armies in the centuries ahead.

Throughout the centuries of Rome's dominance, the army on the march threw a screen of cavalry ahead, both to watch for and locate the enemy and to find sources of water and food. Specialist cavalry units, like the Batavians famed for crossing rivers or the *exploratores*, seasoned scouts, were essential in enemy territory. Training for cavalry work, both on campaign and on the battlefield, was provided at specialist camps, like the newly reconstructed fort at Baginton, Warwickshire, originally built soon after Boudicca's rebellion of AD 60.

The most complete account we have of an invading Roman army comes from Titus Flavius Josephus who was with Vespasian in AD 70 as he rode into Galilee. After the vanguard of experienced, old-hand legionaries, 'rode Vespasian himself, with the cream of his horse and foot and a body of spearmen. Next came the legionary cavalry; for each legion has its own troop of 120 horse. These were followed by the mules that carried the battering-rams and other artillery. After them came the generals ...' The generals led their men, standing out under their scarlet banners. But the cavalry rode ahead of them. Among the *impedimenta* (baggage) carried to the rear of Vespasian's column was spare cavalry equipment, saddles, bridles, girths and horse-dress. With their *alae* would have been the *veterinarii*, the vets to see to horses' wounds and illness. The average cavalry horse eats 20–24lb of hay fodder a day; on campaign, this was probably impossible. Lush grass could provide the seven hours of grazing to make up the daily diet; but how much of that would there be in Galilee? It may have been the land 'flowing with milk and honey', but greenery was in short supply. This was the double importance of the scouting cavalry – travelling up to 40 miles a day, they found forage for their own animals at least, which meant that only the baggage mules needed to be fed.

Historian John Peddie has analysed the diet of the Roman cavalry horse and concludes, rightly, that it was less well-bred and 'fussy' than today's thoroughbreds. He calculates that twenty horses could subsist perfectly well on 800 acres (1¼ square miles) which means that in Gaul during Julius Caesar's wars, there would not have been a problem.

Publius Flavius Vegetius Renatus, the fourth-century military historian whose word was still law 1,000 years later, wrote that the cavalry guarded the camp every night and were relieved every morning because of the fatigue to man and horse. The cavalry horses of Vespasian's *turmae* were tethered face to face along a piquet line of perhaps sixty feet. The tents for the riders, decurions and servants (grooms) were nearby.

By the sixth century, when Rome was tottering, a reversal of roles had occurred in the army. Battles were rare and they were not even fought by legionary infantry, but cavalry, warbands whose speed and mobility could (just) cope with raids that not only probed the frontiers but penetrated into the heartlands. In 408, Alaric and his Goths had sacked Rome itself, promising to take everything except Roman lives as long as the besieged behaved themselves. The military manual *Strategikon*, written in the sixth century, emphasised the need for cavalry to deliver controlled charges to prevent their horses being blown. Prince Rupert was not listening 1,000 years later; neither were the Scots Greys at Waterloo. The book also stressed the need for a cavalry reserve, to have a fresh unit away from the action to deliver a final blow against wavering infantry. These lessons were universal and were born of battle experience from Cannae (against the Carthaginian Hannibal) in 261 BC to Adrianople (against the Goths) in AD 378, both of them Roman defeats.

Where, in all this, is Incitatus? The role of an emperor in war was to lead his armies in battle, although they rarely did. The history books refer to the Claudian invasion of Britain in AD 43, but the actual work was done by Aulus Plautius. Claudius did not even set foot on British soil until the fighting was over. Caligula's foray into military matters was extraordinary. In 39 he led an army of 250,000 across the Alps, travelling in luxury with the (then still loyal) Praetorians at his back. He may have ridden Incitatus and despite the baggage of two of his sisters, made good progress. He launched raids across the Rhine, but the enemy were his own units and these were actually carefully staged war games. He sent reports home to the senate of these 'victories' and at least seven times his soldiers acclaimed him as *imperator*, the greatest accolade as war-leader.

When the prince of a British tribe, Amminus, defected to him, Caligula could claim that he had conquered Britain, giving himself the title Britannicus to add to his other laurels. Camped for months at Lugdunum (Lyons), the high spot of Caligula's British campaign was to order his artillery to fire their catapults into the ocean and collect seashells as booty. Historians still argue about this. Did the army rebel against their emperor, fearing the Channel and what lay beyond? It had not deterred Julius Caesar (twice) and there was regular cross-Channel trade with Britannia, so this explanation, from Suetonius, seems unlikely. It may be that the hard-bitten troops were only too delighted to earn a triumphal return to Rome, with the usual flag-waving celebrations, for having done nothing. They may have grumbled, but most of them could probably live with that.

One account has Caligula galloping hard and fast on Incitatus along a bridge, built by the legions from boats. Another has the animal breaking a leg in a chariot race and having to be put down. Neither story can be verified.

In 2000, BBC radio put on a play *Me and Little Boots*, telling the story of Caligula through the prism of Incitatus. The late Leslie Philips played the white stallion, the most pampered animal in Rome.

Chapter 5

A Gallop Through the Dark Ages

The term 'Dark Ages' seems hopelessly old-fashioned today, implying, as it does, that cultured civilisation came to an end with the fall of Rome. It covers the period from the fifth to the eleventh centuries.

When the Romans left Britain in the late fourth and early fifth centuries, they left behind them a power vacuum where independent warlords ruled over different parts of the country, ultimately divided into a heptarchy of Mercia – broadly Somerset, Gloucestershire and the West Midlands; Sussex; Wessex; Northumbria, (which was divided into two kingdoms); Essex; and Kent – which remained the geopolitical landscape until Alfred and the Viking invasions of the tenth century. It was this final kingdom that gives us the legend of Anglo-Saxon brothers Hengest and Horsa.

'Stallion' and 'Horse' as their names translate from the Old English legend has it, were respected Anglo-Saxon warriors, invited by Vortigern to help deal with invading Picts from the far north. Vortigern was referred to as King of the Britons in many chronicles and histories including the *Anglo-Saxon Chronicle* and Geoffrey of Monmouth's *Annales Cambriae*, but was almost certainly a leading warlord of the time rather than a monarch.

The brothers landed in Kent during the 440s, and quickly dispersed the Pictish threat, then turned their attention to their host in what Geoffrey of Monmouth labelled, in the twelfth century, 'The Night of the Long Knives'. Invited to a celebratory or perhaps political gathering, the Anglo-Saxon mercenaries concealed their seaxes – long bladed knives – under their cloaks, drawing them in the banqueting hall and slaughtering the unsuspecting Britons. Horsa is said to have been killed in battle shortly after their coup, but his brother Hengest ruled Kent for a number of years and is the one from whom subsequent Kentish kings claimed descent, therefore originating the emergence of the Anglo-Saxon heptarchy which continued for the next 500 years.

Whether or not these were real figures, composites of Anglo-Saxon or Jutish invaders, or purely symbolic legends, their legacy endures to this day in 'The White Horse', the emblem of many things Kentish including its police force and the badges of the county's yeomanry regiments.

With all these equine-related symbols and place names, it seems odd that many historians question the existence or at least heavy use of cavalry by the Anglo-Saxons. How were they transported from mainland Europe to Britain? Why do the chroniclers talk mostly of infantry battles, and make specific mention of soldiers dismounting from horses to fight?

These are easily countered, however, by examining these accounts, and the evidence that does exist. One account refers to an expedition into the foothills of North Wales, incredibly unsuitable terrain for a cavalry skirmish, let alone a charge. The Saxons were very proud of, and renowned for, their shield wall, and this may well have been a focus of chroniclers eager to expound its virtues, rather than their ability on horseback. The Bayeux Tapestry, admittedly nearly half a millennium later, shows horses onboard ships. The Norman horsemen are the original knights and were descended from the Jutish and Frankish cultures to which Hengest and Horsa belonged. The Channel is not a massive body of water, so transport, though difficult, was far from impossible. Additionally, there were horses aplenty in Britain already, which could have been utilised by a culture who were familiar with the horse.

Then there is the artefact evidence. One of the most iconic archaeological finds in history, the Sutton Hoo helmet, discovered in a ship burial in the 1930s, has on one of its decorative panels, a horseman riding down an enemy. At Lakenheath in 1997, Colchester Archaeology uncovered a huge Anglo-Saxon burial site at the RAF base there. Amongst the burials was a warrior and in the next pit, a slaughtered horse, his ride into the afterlife.

The Aberlemno stones in Scotland are some of the most varied and detailed pictographic stones in the world, covering depictions from prehistoric cup-marks to the Ringericke and Celtic knot work of the Danelaw. On Aberlemno 2 (of 5), there is the depiction of a battle involving mounted warriors with spears. There is much debate as to what possible conflict this could show, but its context and style suggests that it is mid-ninth century in origin. All of these examples exist in the East of Britain, where Hengest and his brother's domain was established, adding even more weight to the argument that this was a culture of mounted warriors.

> They are ill-fitted to fight on foot, and remain glued to their horses, hardy but ugly beasts, on which they sit like women to perform their everyday business. Buying or selling, eating or drinking, are all done by day and night on horseback and they even bow forward over their beasts' narrow necks to enjoy a deep and dreamy sleep ... Being lightly equipped and very sudden in their movements can deliberately scatter and gallop about at random, inflicting tremendous slaughter.

This less than flattering but ultimately revealing account from Ammianus Marcellinus, a fourth-century Roman chronicler, refers to one of the most feared groups in early history, the Huns.

These nomadic horsemen had been used extensively by the Romans as mercenaries against other tribes such as the Visigoths, but by the opening years of the fifth century, they had begun to wage war on all fronts and were allied to none. This led to an almost unstoppable wave of Hunnic attacks East and West which ravaged the known world for nearly half a century. They are quite an enigma to the archaeological record, but their successes and reputation were such that their existence and appearance could not help but be well documented in contemporary histories.

The key to the Huns' success was their relationship with their horses. The tribes of Central Asia and the Steppe were hugely reliant and attached to their horses, and were literally 'born in the saddle', as most were nomadic, and their horses, quick and tough, were essential to survival in a tribal landscape. The majority of these horses were also mares, which meant they could also sustain the life of their rider though their milk, which was plentiful, as they were able to graze in even the most harsh winter conditions. Steppe ponies, which are still in use, are small, around 12 hands 2 inches (1.2m) high, muscular and agile. Attacks were fast, confident and furious, and made use of melee and skirmishing tactics well ahead of their time. Before the invention of gunpowder, there really was no force that could stop these mounted warriors. It was their 'otherness' that proved so formidable on the battlefield. The Hun were confident, as they lived and died in the saddle and were not tied to the strictures of sedentary and centralised government. On the battlefield, the use of mounted archers was unique. Although they were also renowned for their spearmen, in general riders would not charge head on into the enemy, a tactic that was used by practically every cavalry unit before and after, but instead would encircle and ride around their enemy, loosing barbed arrow after barbed arrow from small composite bows to devastating effect in a scene similar to so many Western films. It was only with the death of their greatest leader Attila in 453 that internal divisions led to their power and effectiveness being weakened. But this was not before the Huns had amassed a vast fortune, and also begun to settle large parts of Central Europe, being joined by many other horse soldiers. The Alans, descendants of the Samartians, and therefore Iranian in descent, had made a similar journey into Central Asia and the Steppe like many of their equestrian counterparts. With already strong pedigree, the Alans' horses were equipped differently from the Huns, whose light armour and tack made them so agile in the field. The Alans were more direct, using long swords with both hands, and covering themselves and their horses in heavy leather and metal hauberks. They were both allies and enemies of Hun and Goth, Roman and Scythian, but were always very significant in political decision making in Central Europe in the Dark Ages. One of their most tenacious enemies also provide one of the clearest examples of a huge advancement in equestrian warfare.

Probably the single most important development since the rein really took hold in the opening centuries of the Dark Ages, the stirrup has been identified as early as the second century BC in India, although its use as a military development does not really become apparent until the sixth century AD. The Avars were a group of nomadic tribes originating in Northern China, a subset of the Huns. Like many cultures from Asia and the Steppe, the Avars were skilled horsemen. Horses were very important in this harsh landscape, as they were able to survive frozen conditions and cope with rugged terrain, being domesticated but largely the same as their ancestors in terms of size and agility. The Avars seem to have adopted the stirrup from other Central Asian groups such as the Parthians and were early adopters of the stirrup as this gave them an ability to fire, and to support themselves. The Parthians had, over their 500-year history *c.* 200 BC to AD 250, warred with Rome and China, who may well have taken the stirrup from them, adding to the debate as to whether the stirrup derives from China or the Middle East. It is also possible that the Avars adopted the stirrup from China, as there is archaeological evidence for Chinese stirrups dating from the fourth century AD.

By the mid-sixth century, the Avars had migrated to Central Europe and established a Khanganate in Hungary, Romania, Serbia and parts of Turkey under their greatest leader, Bayan I (AD 562–602). They had formed a fragile alliance with the Byzantines, who used them successfully to displace a number of other tribes in parts of Central and Eastern Europe, such as the Gepids. But by the early seventh century, the Avars were not content with being Byzantine's mercenaries and launched an ultimately fruitless attack on the centre of the Byzantine Empire in and around Constantinople itself. One thing that the Byzantine Empire took from the Avars was the stirrup, which is recorded in a Byzantine military manual dating from AD 580, shortly after the Avars had gained control of large parts of an area known as the Pannonia or Carpathian Basin.

With the Byzantine Empire now utilising the stirrup, it quickly spread throughout Europe and soon cavalry units were able to enact more devastating attacks, as the stirrup not only provided control of the horse, but also support for a rider wielding increasingly longer spears and lances. By the end of the Dark Ages, the stirrup was also instrumental in the development of a new class of warrior, that of the knight, as it gave the ability to support a fully armoured rider.

This period in history is very much centred on the horsemen of the Steppe, or their influence at the very least, but it is important to make mention of a group not really associated, perhaps incorrectly, with the horse. The Vikings of Scandinavia were in fact one of the first after Byzantium to adopt the stirrup, but their use of the horse is not well documented or evidenced in archaeology. We do know that they had their own breed of strong hardy ponies, which still exist in Iceland and Shetland, but as far as history allows, the use of the horse was largely for scouting or travelling to battle, whereafter they would dismount to fight. The argument for transport of horses is more valid here than in the Anglo-Saxon case, as many of their raiding ships were deliberately small to navigate internal waterways, and therefore too small to transport horses. The Iceland and Shetland examples are the exception rather than the rule as these were specifically expeditionary and not raiding-based journeys and ships were larger. But that is not to say they did not steal horses once on land, and the hypothetical evidence for this comes in the shape of probably the first great heavy cavalry, and descendants of the Vikings, the Normans.

Chapter 6

Babieca, the Booby

Odo of Bayeux sat his horse under his lord's leopard banner, confident that the pope's gonfalon fluttered there too. Yes, he was a bishop, but God's work today would be of a particularly bloody kind and in case any moralists complained of a churchman taking a life, he carried no sword. Instead, his weapon of choice was a two and a half foot mace of oak studded with iron. Odo's tonsure was not visible today; it was shrouded in a coif of mail under a conical iron helmet, his face obscured by the broad nasal that protected his nose.

It was 14 October in the year of his Lord 1066, St Calistus' Day. We have no idea of the weather that day, but in mid-October, dawn is cold and the grass heavy with dew. The Norman army, only recently arrived after a grim crossing of the Channel, was drawn up in battle array at the foot of Senlac Hill, about 7 miles from the coast where they had arrived with their prefabricated wooden castle. William of Mortain sat next to Odo and behind him, the flower of Norman cavalry, in shin-length mail hauberks and carrying kite-shaped shields, were led by tried and tested warriors like William Malet and Roger le Bigod.

To Odo's right, Eustace of Boulogne's and Robert of Montgomery's banners flapped at the head of their French mercenaries, the Flemings and the Picards. To his left sat the men of Brittany under Count Alan Fergeant and Neal of Saint-Saviour.

Odo may have found himself chuckling. The English shield wall, at the top of the hill, may have looked formidable, extending perhaps a mile wide, but there was no cavalry. The English leaders had ridden to the field, but then dismounted and prepared to fight on foot, as they had done for centuries. This was going to be a walkover.

For much of the history of the last thousand years, the myth of the battle that came to be known as Hastings was that the new tactic of the cavalry charge swept away the old infantry formation of static warfare, delivered at the speed of a man on foot. This is an exaggeration. As things turned out on that October day at Senlac, William of Normandy's repeated cavalry charges won the day, but only after several hours fighting and the desperate use of both infantry and archers. Nothing else could explain why, throughout the Middle Ages, the English aristocracy continued to fight on foot while the French mounted (increasingly futile) cavalry charges. When delivered properly, a cavalry charge was irresistible, few infantrymen being willing to stand their ground against it. When carefully countered, however, as by Robert the Bruce at Bannockburn in 1314, they could be violently self-destructive. The Bruce had very few horsemen and on the night before the battle, dug deep pits with sharpened stakes at the bottom, covering them with brushwood. The English nobility charged and the result was devastation. Recent interpretations of this battle have played down the importance of these 'pottes', but, coupled with boggy ground around the tributaries of the River Forth, they were vital.

At Hastings, Norman archers opened proceedings, but their arrows bit uselessly into the limewood shields of the English housecarls and fyrdmen and William sent in his infantry. Harold Godwinson's shield wall held firm against several attacks, the men of the housecarls swinging their two-headed axes which could literally split a man's head. The Bretons fell back and over-exuberant fyrdmen broke formation to chase them. William's cavalry retaliated and killed most of them on a hillock halfway down the slope.

At one point, the rumour spread – and rumours on a battlefield are poison in terms of the psychology of a fighting man – that William was dead. He rode in front of his wavering men, tilting back his helmet and roared at them in the harsh voice they knew all too well – 'Look at me well. I am still alive and by the grace of God I shall yet prove victor.' He shook his mace at them and forced them back into formation. In the end, the sheer momentum of the Norman cavalry carried the day. Harold Godwinson's banners of the dragon of Wessex and the Fighting Man were clear targets and, with the dead of the shield wall held upright by the living, the king himself was cut down, probably already wounded with an arrow in the face.

'Dex aie' (God help us) the Normans roared and the equally defiant battle cry of the English – 'Out! Out!' – died away as the shield wall finally broke and men limped away over the rough ground that came to be known as Malfosse, the evil ditch, and made for the woods. Although no one who fought there that day knew it, a way of life had come to an end and there would be a foreign king on the throne in a new capital; for Harold, elected by the Witan, the council of England, read William, a usurper on the throne by force of arms. For Winchester, the ancient capital of Wessex, read London, a frontier town famous for its foreigners.

What do we know about the horses at Hastings? Much of our information, at least visually, comes from the Bayeux Tapestry, a unique piece of Romanesque appliqué work (and not a tapestry at all) which may have been commissioned by Bishop Odo of Bayeux, who we know fought at Hastings. But if Odo was there, the anonymous nuns who designed and made the work were not and eleventh-century artwork is as unreliable as anything the Assyrians,

Egyptians, Greeks and Romans produced, particularly when it came to horse depictions.

The Tapestry, which has a colourful history of its own, was probably made about 1077 when both Odo and William were still alive. It is a sort of 'graphic novel' covering a period from 1064 to October 1066 and the action usually moves from left to right between borders with abstract designs and mythical beasts. The dozens of horses are all depicted in the same way; their legs furthest from the viewer are in a different colour from the nearside legs and body. This is probably an attempt at shading. There is no difference between, say, the palfreys and rounceys ridden by Harold on his hunting expedition and the warhorses ridden by William's men at Hastings yet, in reality, the warhorse was taller and heavier than the everyday animals. The only word stitched on the Tapestry in connection with horses is *caballi*, which is not strictly Latin, hinting perhaps that it was *English* nuns who made it.

Over the years, experts have studied the Bayeux Tapestry's horses in the context of harness, horsemanship and breeds. One of these, Lucien Musset, writing in 2002, believes the horses at Hastings were 'not of great height and archaeological evidence confirms that animals of this era were more the size of large ponies than of today's carthorses'. Others believe that the Norman animals were destriers, the 15 or 16 hand precursors of the shire horses, who we know were being bred in the thirteenth century. Given the nature of the ground at Hastings – the relatively steep slope of Senlac and a marshy area around a stream – William's cavalry charges were probably delivered at a trot rather than a gallop, so we have no way of gauging the Norman horses' weight. If they weighed, as some experts maintain, a ton and if William had 3,000 of them, the impact on the Saxon shield wall must have been incredible.

The Tapestry shows all Norman horses as stallions, but, as we have already noted, this probably had more to do with the accepted wisdom of male egos than practicality. Geldings (neutered stallions) or mares were probably more common as being more reliable. The stirrups worn are triangular, very like the contemporary iron version overlaid with bronze in the Kunsthistorisches Museum in Vienna. The saddles have high pommels and cantles, providing a supportive seat for the rider much like the Western American saddle of cowboy fame. This is important because it enabled the Norman knights to wear their stirrups long and effectively 'stand' in the saddle to deliver their lance blows with the weapon tucked under the arm rather than thrown as a javelin. A saddle of this type played its part in the death of William of Normandy. While laying siege to Nantes in 1087, the king's horse leapt a ditch and the pommel rammed into William's chest, causing a wound which led to peritonitis and death.

In the Tapestry, the Norman horses are saddled and bridled for their stormy crossing of the Channel. There is no mention from any contemporary as to how this went but transporting horses in rough seas remained a problem for armies for centuries. Unused to the wobble of a ship and penned together in cramped, often dark conditions, animals panicked and lashed out with their hoofs, hurting each other and smashing ships' timbers with potentially fatal results. Although it bears little resemblance to what must actually have happened, the Tapestry shows the horses being disembarked at Pevensey – 'here the horses leave the ships'.

The horses' manes are in all cases unbraided (unlike most of the ancient world's examples) and while there is a breast strap, there is no crupper to secure the saddle. The reins are single and the bridles simple halter-type, although this may be a simplification on the part of the embroiderers.

We know that William had two horses given to him as a gesture of goodwill by Alfonso the Brave of Spain. These were probably Andalusians, the powerful breed developed from North African Arabs and native Spanish animals. We do not know whether the duke rode either of them at Senlac. Andalusians were easily distinguishable with their proud, arched necks, thick mane and tails. Their average height was 15 hands 2 inches and the men who bred them believed that particular markings were a sign of good or bad luck. White stockings could go either way, but a horse without white markings of any kind was considered likely to be bad-tempered.

About twenty-five years before the Norman conquest of England, a son was born to Diego Lainez, a hidalgo of Bivar, near Burgos in Castile and a woman whose name is not recorded. The boy was christened Rodrigo and as an adult he acquired a military reputation second to none. Men called him *el cid campeador*, the lord champion, and he owned a warhorse called Babieca, the booby.

Eleventh-century Spain was a land torn apart. In the north, dominated by the kingdoms of Castile, Leon, Aragon and Asturias, Christianity was paramount. The area was little different from France to the north and the surrounding duchies, personified by Normandy. Mid and southern Spain, however, was Al-Andalus, where mosques stood in place of churches and the crescent flags of Islam flew over the great palaces of Granada, Seville and Cordova.

The Muslims had crossed the straits, which had been called the Pillars of Hercules in the ancient world, in 711. Led by a formidable general, Tariq ibn Zayid, they took the peak over the harbour and named it Tariq's mountain – Gibraltar. Four years later, the Moors, as the North African invaders were known, had crossed the Pyrenees and had pitched their tents in what is today the south of

France. It would take 500 years of *Reconquista* (reconquest) before the Christians had won back all Spain.

Rodrigo Diaz reached maturity during the reign of Fernando of Castile. We know that his sons, Sancho, Alfonso and Garcia, were taught to ride 'after the manner of the Spaniards, to practise the use of weapons and to hunt', according to the contemporary *Historia Silense*. This riding style, with long stirrup leathers and the lance couched under the arm, was different from the tradition of Al-Andalus, where riders used short stirrup leathers and rode with the knee bent (today's technique). The future Cid Campeador would have received similar training. He reached the royal palace of Leon in 1057 when he was about seventeen.

According to a family legend, Rodrigo had already acquired his famous horse by this time, but the story is too reminiscent of Alexander and Bucephalus to accept wholesale. The boy's godfather was a priest known behind his back as Peyre Pringo, Fat Pete. The indulgent friar told his godson to select any of a herd of horses he owned and the lad chose an ugly, scrawny colt – 'This is the one for me' and Pringo, horrified at the boy's ignorance of horseflesh, called him *babieca* (idiot). The name stuck, not to Rodrigo but to his horse. The story comes from the fifteenth century, by which time a number of literary efforts had been produced, chronicling the life of the Cid, which strayed ever further from the facts. Artistic depictions of the Campeador and his horse do not exist during their lifetimes. A typical woodcut of 1512 from the *Cronica particular del Cid* shows a surprisingly small warhorse, perhaps 14 hands, but since Rodrigo is also shown in sixteenth-century plate armour, this is not much of a guide. The assumption is that Babieca was an Andalusian destrier, the heavy horses bred for battle throughout the Middle Ages, although, as we have seen, this convention has been challenged by recent experts.

From these later accounts of the Cid's life, we have the hero defeating five emirs before he was 20 and Babieca was 3 or 4 years old. When Rodrigo captured Valencia, far to the south in Muslim territory in 1094, horse and rider became famous throughout Spain. The story goes that Rodrigo, out of loyalty to King Alfonso (the Brave), offered him the horse. 'God forbid that I should take him,' Alfonso replied. 'A horse like Babieca deserves no other rider than you, my Cid, so that both together you may drive the Moors from the field …'

The death of the Cid featured Babieca too. Wounded in defence of Valencia, Rodrigo died during the night, but the next day, he appeared, strapped on Babieca's back, galloping through the lines of astonished, terrified Moors.

> The body, thus as it was,
> They placed upon Babieca
> And onto the horse tied it.
> Erect and upright it sits,
> It looked as though it were living …
> On one side rode his bishop.
> The famous Don Jerome,
> On his other Gil Diaz
> Who guided Babieca.

Macabre and magnificent though it is (the above translation is by the English poet Robert Southey from the fifteenth-century chronicles) none of it is true. The real Rodrigo died of natural causes, not in battle, and his widow, Ximene, took the body back to Burgos for burial. The *Chronicle of the Cid* covers the last years of Babieca too, who outlived his master by two and a half years (he would have been 40 by now). No one rode Babieca after the Cid's death and he sired a race of magnificent animals in Castile. He was buried outside the gate of the monastery of San Pedro de Cardẽna. The grave was excavated in 1948 but no remains were found. Even so, the Duke of Alva had a memorial dedicated to the horse, to join various statues that exist in some very unlikely places around the world, from Seville to New York.

The archetypal Medieval warhorse was the destrier or 'great horse'. The term comes from the Norman-French *dextrarius*, right-handed, although the exact meaning of this is disputed. Did riders lead their mounts by the right hand, on the animal's near side? Or does it imply strength, in that fighting men were trained to use their right arms, which became stronger with practice? The destrier was not a breed, but a type, although from the conqueror's time, Spanish and French breeds were imported to improve the size and stamina of warhorses. Some of the earliest recorders of horse pedigrees, from the thirteenth century, were Carthusian monks, used to keeping books of all kinds in their monasteries. William Fitzstephen, secretary to Thomas Becket, Archbishop of Canterbury, described the horse fair which was held at Smithfield, outside London's walls every Friday. 'It is a joy to see,' he wrote, 'the costly destriers of graceful form and goodly stature with quivering ears, high necks and plump buttocks ... When a race is about to begin ... their limbs tremble, impatient of delay; they cannot stand still. When the signal is given, they stretch forth their limbs [as] they gallop away ...'

If Fitzstephen is referring to destriers here, they put on weight over the next 200 years! Edward II's knights at Bannockburn could manage little more than a trot. These were the horses, perhaps 16 hands, that Richard the Lionheart took on crusade in 1189. One ship carried forty horses and sufficient forage to feed them for up to a year. The Cid died in 1099, the year of the First Crusade, the only truly successful one from the West's point of view and this brought the Medieval knight and the destrier up against a different breed – the Arab.

One of the many apocryphal tales of the crusades is the contest between the Lionheart and his opponent, Salah Nagin al Din, to prove the craftsmanship of their respective swords. Richard, an accomplished warrior six feet tall, smashed an anvil with his broadsword. Salah Nagin al Din, by contrast, a philosopher and doctor of medicine, threw a silk handkerchief into the air and let it land on the blade of his scimitar, which was so sharp that it cut the cloth in half. Similar nonsense is found in comparing the huge, lumbering 'carthorses' of the crusades with the fleet-footed Arabs of the Muslims.

'The Arabs' wrote the poet Ibn Rashing in 1064, 'congratulate each other on three occasions: the birth of a boy; the rise of a poet in their midst and a mare giving birth to a foal.' The Arab horse was originally the property of the nomadic Bedouins and was endowed with phenomenal stories of speed, agility and grace. Such animals lived on air and held the same status as horses in Celtic mythology long before the coming of Islam in the seventh century. The hard geography of the Arabian Peninsula produced little water and less food so the horses there had to be hardy. The climate made them resilient, strong and they were believed to be particularly intelligent. In Arab mythology, a man's blood belonged to his tribe, his soul to Allah and his heart to his horse. The animals slept in the Bedouins' tents and were treated as part of the family; their genealogy was recited like an Islamic mantra.

The origin of the Arabs is in dispute. Many believe that the first identifiable animals were the five horses of Sulaiman bin Daud – the biblical Solomon, whose stables in Jerusalem were commandeered by the Knights Templar after the First Crusade. From these, five distinctive sub-breeds developed, characterised by colour and temperament. Most obvious perhaps were the Saqlawiehs (from the Arabic *Saqla*, gallop) with a dancing gait and the spoon-shaped skull and upturned nose that is typical of Arab horses.

Because of their docility, mares were preferred in battle to stallions. The prophet Mohammed rode Jellabiya to commemorate his victories in the early spread of Islam. The Koran, which is essentially the wisdom of Mohammed mixed with poetry, claims that the horse is Allah's tool and that it was the duty of Muslims to breed pure horses in Allah's name. Interestingly, there is no similar tradition in Christianity, no matter how fond a crusader might be of his mount. The only animal that Christ rode was an ass, regarded as an inferior animal, a beast of burden, which is precisely why Jesus rode him. In Mohammed's seventh century, owners of pure-bred Arabs received seven times the war booty given to the infantry and twice as much as the riders of non-Arabs.

'If a man cannot fulfil all his religious duties,' says the Koran, 'then let him keep an Asile [pure-bred] horse in honour of Allah and all his sins will be forgiven ... He who has bred an Asile for jihad [holy war] will be saved from Purgatory on the Day of Judgement.'

The Arab horse was part of Muslim propaganda from the first – Allah made the animal from the south wind – 'I have granted you the power to fly without wings whether in attack or retreat. I want to set men on your back who glorify and praise me and sing Hallelujah ...' Allah's orders to the horse were 'Humiliate the idolators with your neighing and fill their ears with it and fill their hearts with fear.'

Most Arabs are greys and the Obeyah bloodline has the characteristic of holding the tail high when galloping. Because of the grey preponderance, black horses were particularly prized by the Bedouin. Most horses wore turquoise chains around their necks to ward off the evil eye and the strictest Muslims would not show their finest horses to infidels; on the battlefield, of course, it was different. It is no exaggeration to say that the astonishing speed of the spread of Islam was due to the Arab horse. No Western poets extolled the European horse, but Arab poets constantly sang the praises of their bloodstock:

> Lo, the mares we bestride at the dawn of battle!
> Sleek coat mares, the choice ones; we have reared them.
> Charge they mail-coated together, how red with battle,
> Red the knots of their veins as dyed with blood.
> Are not these the inheritance of our fathers?
> Shall we not to our sons in turn bequeath them?

This was Amir bin Kulthun about 530, prefiguring the importance of the horse in Muslim warfare. Mohammed himself was not a natural cavalryman – his caravans relied on camels instead, but he lost his first battle against the Meccans in 624 because their cavalry were better than his and he learned from it.

The Muslim cavalry of the crusades was made up entirely of mares – stallions were unpredictable and gelding was contrary to the will of Allah (except in the case of human eunuchs). The knights, largely from France, who followed Godfrey de Bouillon to Outremer (Palestine) in 1099, would not have looked very different from William's men at Senlac, except that they wore gold or red crosses on their surcoats. The intense heat of the Middle East led to tying cloth around their helmets to keep the sun off their necks, which in time morphed into the fanciful mantling of heraldic decoration. Those men, enthused by the promise of salvation from Pope Urban II and exempt in the eyes of God for the murder of Muslims, were already the elite, landowners at the heart of the feudal system, serving their king where and when he commanded. To this end, they were obliged to bring their weapons and horses in royal service. In reality, many of them were thugs, that part of Medieval society called *pugnari*, those who fight. The concept of chivalry which centred on them, with its notions of duty, courtesy and valour, were rarely echoed in reality.

After de Bouillon's success in Jerusalem in 1099, during which his men sloshed in blood up to their prick-spurs, the new kingdom of Christ had to be monitored and guarded and a number of military Orders, part knight, part monk, were set up to keep the cross flying over a city that everybody wanted. The most famous of these were the black-flagged Hospitallers and the red-crossed Templars.

The latter's headquarters was the Temple of Solomon, the famous horse-breeder of the Old Testament and the Templars used its outbuildings to stable their own horses. The symbol of the Templars, to emphasise their poverty (a noble trait they quickly dropped!) was two knights riding one horse. Years later, when the excessively rich and powerful Order had come into disrepute, cynics pointed to this as symbolic of pederasty, then an abomination punishable by death. The daily routine of the Templars, which has survived, makes several references to the need to feed, water and groom the horses as a task as important as prayer.

In typical battle formation, echoing the Greeks and Romans, the infantry formed the centre with cavalry on the wings and there is no doubt that the Franks (as crusaders were called by their enemies) relied on the shock charge of these horsemen to rout opposing infantry and scatter them. The problem was that the lighter-armoured Seljuks and Mamelukes they faced did not play by the same rules. Like the Parthians of the ancient world, they used mounted archers to great effect, using hit-and-run tactics which literally ran rings around the heavier crusader horsemen. For over a century in eighteenth- and nineteenth-century Britain, experts argued about the best shape for a cavalry sword – straight or curved, but the Muslims invariably used the curved blade and this was usually superior.

By the late twelfth century, the Mamelukes dominated in Egypt, about 10,000 of them, trained in archery, offering their services to Saladin. So successful did these troops become, retaking crusader bases at Jaffa and Antioch and driving the Mongols and Turks out of the area, that they became rulers of Egypt by the fifteenth century.

The crusaders learned a great deal from the crusades, especially in the art of siege warfare and artillery technology. The great castles built by Edward I to subdue Wales in the thirteenth century owe much to the work of Arab architects in the Middle East. In terms of cavalry warfare, however, most lessons were ignored. The Holy Land was a unique place in terms of geography and climate; fighting in Europe was different, so the old ways were retained.

Central to this was the training given to men and horses. Teenaged squires rode first the palfreys or rounceys, the small riding horses of the day. They then progressed to the destrier, legs straight, saddle double-girthed and rode at the quintain, striking a shield attached to a pivot and moving quickly enough to avoid being smacked around the back of the head by the morning star (mace) that hung from the pivot's other arm. Although knights faced each other, usually on feast days or to celebrate coronations and weddings, in the lists, single combat using lance and shield, the more common form was the melee. This was a free-for-all of teams of knights, up to forty men in all, bashing each other in a sort of armed rugby match. *The* expert in England was William Marshal, military adviser to Henry II and his sons Richard and John. Such fights were spectacular to watch and useful practice for the real thing, but they rarely allowed individual skill to shine through and were dropped in the later Middle Ages.

There was no armour specifically for the horse in the early Middle Ages, although various depictions in thirteenth-century manuscripts show bards of mail which cover the animal's body in the same way as heraldic cloth coverings did. The Luttrell Psalter shows its owner, Sir Geoffrey Luttrell, being armed by his wife and her lady-in-waiting. His grey destrier is covered with an heraldic cloth bard with his arms of the bend and birds. The animal's head, complete with a comb crest similar to that on the knight's great helm, has an iron chamfron on the forehead held in place with leather straps. But this is from about 1340, at the start of the Hundred Years War and plate armour was rapidly developing in this period.

Actual examples of horse behaviour in Medieval battle are rare. Art invariably shows animals galloping, with all legs off the ground (a physical impossibility except for a split second) which would not finally disappear from horse painting until Eadweard Muybridge's famous series of photographic stills from 1872. One of the rare documented accounts of a horsed duel took place at Bannockburn when Robert the Bruce took on the English knight Henry de Bohun. De Bohun was the cocky young nephew of the Earl of Hereford, hoping to make a name for himself in individual combat. He had his chance when he came across Bruce, recognisable by the gold circlet on his helmet, riding, according to a contemporary, 'on a grey palfrey, littill and joly'. De Bohun charged on his destrier but Bruce wheeled the pony out of the way and smashed the Englishman with his battle-axe 'to the brisket', complaining afterwards that he had ruined his best weapon. It was almost an action foreplay of what was to follow for Edward II's entire army.

Chapter 7

The Wind Horse

Subedei Bat'atur of the Reindeer People sat on his little pony in front of the great stockade his troops had built. Before him stood the walls of the Russian city of Riazan, flags of their Christian God fluttering in the breeze. His huge army had accomplished wonders already, crossing the frozen Volga in midwinter, riding through the dense pine forests like ghosts. If anybody in the few scattered villages they passed caught a glimpse of them, they ran. The Mongols were not men; they were monsters, more terrifying than any of the bogatyr, the gods who haunted the forests and fens. No one could stand against them.

Subedei was not a monster. He had tried diplomacy. He had sent a woman ambassador with two outriders to the gates of Riazan. She spoke their language, although the Russians had been appalled to see a *woman* trying to negotiate with them; if that did not prove the barbarism of the Mongols, nothing would. Subedei of the Reindeer People knew that that would not work. He had kept his hordes of horsemen miles back in the forests, bringing his engineers forward to build their stockade.

All around the city it ran, ten foot high at its lowest, standing timber to timber with the wooden walls of Riazan. Subedei had asked the prince for surrender, a tax of 10 per cent and reinforcements for his troops; not unreasonable in the scheme of things. The reply had come that once the inhabitants had been allowed to leave, the city was there for the Mongols to take. Subedei knew that this was the Russian way – avoid battle and let their winters do the fighting for them. That was all well and good for the soft soldiers of the West, the Teutonic knights and the Poles. But the Mongols were born in snow; they dipped their newborn babies in the ice.

For five days, the terrified citizens of Riazan had listened to the clatter of hammers and the grunt of men hauling timbers, the rasp of saws and the dull echo of wood being locked together.

Subedei was ready. He had watched the ramparts carefully, waiting for some response – the white flag that meant surrender. Nothing. Well, he had waited long enough. Behind him, the standard bearer raised the banner of the nine yaks' tails. The Russians probably would not understand it, but Subedei's people did. War to the death. No one would survive, unless Subedei decided they should.

He roared the command and all around the silent city, the Mongol palisades came crashing down. The watchers on the ramparts gasped in horror. Every few yards apart, giant ballistae faced them, the technology stolen from those inscrutable bastards, the Chin. At another bellowed command, the huge bows fired, their horsehair cords shuddering as the iron-tipped bolts smashed through Riazan's walls, splintering timbers and scattering the defenders. Man after man fell back from the ramparts, impaled by the arrows, their shields split, their bodies holed.

Then the real bombardment started. Catapults sent flaming clay tablets hurtling through the air, high over Subedei's head. On impact, they exploded; another Chin stratagem. Men and walls disintegrated. Dozens of these grenades hit Riazan's gates and the huge timbers cracked and broke, bursting into flames, yawning open to the city's tangle of streets.

Subedei whirled his sword arm in the air and his cavalry moved forward, clattering over the fallen palisade screen, galloping over the open ground before Riazan. Yellow men on little horses swarmed over the city like ants, leaping the smouldering gates, hacking at anything in their way. They cantered across the squares, batting aside the garrison who tried to stop them. They trotted down the narrow alleyways, scything anything that moved. Women, children, cats and dogs, whatever came their way was fodder for the Mongol swords.

When it was all over, Subedei rode his cream-coloured horse into the royal palace itself, up the steps to the throne room. His men lined the walls and the prince of Riazan stood with his family in a huddle. The Mongol commander never left the saddle. His horse had carried him through miles of enemy territory, through battles without number. His horse made men like Subedei possible. And now, his horse would witness the end of five long years of campaigning.

Through his female interpreter, he let the prince know what would become of his city, his family and his people. The women, even the nuns – especially the nuns – would be raped in front of their menfolk. Then they would be sold into slavery. The prince and his family would be impaled, skewered on poles and left to rot as food for the ravens and the vultures. Riazan would be burned to the ground, as if it had never been.

Subedei half turned to the woman beside him. 'Tell them,' he said, in the strange, guttural language the conquered did not understand, 'I will spare one man in twenty, that they may go and tell the Russians that the Mongols are coming.'

The traveller Elizabeth Kendall, riding through Mongolia in 1911, wrote:

> To appreciate the Mongol you must see him on horseback – and indeed you rarely see him otherwise, for he does not put his foot to the ground if he can help it. The Mongol without his pony is only half a Mongol, but with his pony he is as good as two men.

The Mongolian Steppelands are a perfect breeding ground for horses. Today's horse population outnumbers humans. Unlike Western animals, brought up to a diet of oats and barley, difficult to acquire on campaign, the Mongolian ponies are tough, surviving on the coarse grass of their homeland. They roam freely without stabling in temperatures of 30 degrees Celsius in summer to minus 40 in winter. In terms of modern breeding categories, the Mongolian wild horse was 'discovered' by Captain Przewalski in 1878 on the western edge of the Gobi Desert. The animals had changed little from the wild horses of the last Ice Age and were noted for their cream colour, their short-legged stature (12 to 14 hands) and their vicious temperament. They were herded rather as American cowboys corralled their cattle in the late nineteenth century, except that the Mongols were a nomadic people who wandered the Steppes at will.

The Mongols do not name their horses, but they have a strong affection for them and they are central to historic Mongol economy. Horses were used for meat, and mares' milk, *airag*, is still a national drink. The animals went unshod, which saved time and money. While the Teutonic knights and princes of Novgorod worried about blacksmiths, farriers and the whole labour-intensive business of shoeing, the Mongol hordes could get on with what they did best – plundering.

By Western standards, Mongolian care of horses was almost non-existent, but a sickly foal might well be taken into a ger to be kept warm by the fire. The job of herdsman was mostly carried out by the stallion. 'Each drove of horses,' Elizabeth Kendall wrote, 'is in the charge of a stallion which looks sharply after the mares, fighting savagely with any other stallion which attempts to join the herd … the owner only needs to count his stallions to be sure that all the mares have come home.'

Horses were given as valuable presents at weddings and there was a ritual about the animals that has continued into the twenty-first century. The date of castration of a colt was established by the local lama (priest) and one of the testicles was eaten by the head of the household to give him strength. In horse racing, the winners are patted and stroked by the crowd as a token of respect (as in the West) but also to ensure good luck for the humans concerned.

Mongolian horsemanship surpassed anything seen in the West for centuries. Subedei's archers fired their short bows from the saddle, sliding sideways and shooting from below their horses' necks, as the Parthians had done in the ancient world. When Giovanni de Carpini, a Franciscan friar intent on missionary work, visited Mongolia in the 1240s, a few years after Subedei destroyed Riazan, he discovered that Mongol children were taught to ride at 2 or 3 years old. 'They are given bows to suit their stature and are taught to shoot; they are extremely agile and intrepid. Young girls and women ride and gallop on horseback ... like men.' No doubt, the friar was horrified by all this. He was probably surprised that Mongol horsemen did not use spurs, but a short whip. This hung permanently from a leather loop around the rider's wrist and it must not be touched, according to Mongol tradition, by an arrow. In the peculiar, but in some ways enlightened, legal code, the Yasa, drawn up by the Mongols in the early thirteenth century, striking a horse with a bridle was an offence punishable by death.

The greatest leader of the Mongols was Temujin, probably born in 1167 in the Bjorgin clan, one of the many tribes of the Steppes. His father was murdered when he was nine and Temujin's entire life was spent in warfare against neighbouring tribes and foreigners, like the Tartars, the Chin and, ultimately, the West. In 1204, the Year of the Rat in Chinese mythology, Temujin destroyed his last internal enemy, the Namans, and a *quiriltai*, a summit meeting of tribal leaders, declared him Lord of the Steppes and the perfect warrior – Genghis Khan.

The new ruler, backed by sons who were all competent generals and, perhaps the most able general of his time, Subedei Bat'atur, set up a *nokar* system, similar to the West's feudalism, but based on booty obtained from conquest rather than land. He also created the *keshig*, a Praetorian guard of cavalrymen numbering 10,000 after 1206. The *turghaut* (day watch) was 1,000 strong, the night equivalent (*kabstaut*) the same. One thousand men made up sharpshooters, men skilled with the horn and wood bow and carrying up to sixty arrows in a quiver. The khan's personal bodyguard (*baataut*) was 7,000 strong, so that he was unreachable on the battlefield.

Like the Ottoman army of three centuries later, the Mongols were a society dedicated to war. Seven hundred years before Revolutionary France invented conscription, the *levee en masse*, every Mongol man aged 14 became a soldier for the great khan. Bred to the saddle and to the hunt, these boys became men very quickly, pursuing their quarry, be it game or enemies, for miles in complete silence. There was no baggage train with a Mongol army. Each warrior carried his sword, his bow, his lance, food bowl and a waterproof bag for fording rivers. He also led two or three spare horses and could leap, in his short stirrups, from saddle to saddle in the event of a horse tiring or being killed. The typical Mongol tactic of open battle was to advance in a long single line formation at the walk, gathering pace. The ends of the line increased their speed, forming pincers to encircle the enemy who had usually never seen anything like this before.

Behind the horsemen came the camp followers, family members, also mounted, who acted as cooks and nurses. The gers were loaded on to horses' backs ready-built and could be tent-pegged in moments.

Genghis Khan's cavalry wore silk shirts – cheap in the East – which reduced the impact of enemy arrows. Over this, a tunic of thick felt acted as an extra defence under the breast and back plates of overlapping leather and iron plates. Unlike Western knights, whose great helms restricted sight, hearing and even breathing, the Mongols' faces were exposed, their heads protected with iron or copper helmets. Their leather-covered wicker shields could be slung over their backs, leaving both hands free for combat. On the battlefield, signals were passed from flank to flank with flags and gallopers and at night with flaming torches.

The great khan's orders to Subedei were simple enough:

> See to it that your men keep their crupper [tail harness] hanging loose on their mounts and the bit of the bridle out of the mouth, except when you allow them to hunt. That way they won't be able to gallop off at their whim. Having established these rules, see to it that you seize and beat any man who breaks them. Any man who ignores this decree, cut off his head where he stands.

The Medieval Mongol saddle was flat, without high pommel or cantle, to allow a rider to turn and swivel, to use his sword or lance and fire his bow. The high-arched saddle of Western knights, especially at the tournament, allowed no such freedom of movement. The short stirrups allowed a horseman to stand in them to deliver his attack with maximum force.

Modern Mongol horses – and this was probably true of their Medieval ancestors – are amazingly strong for their size. Experiments have shown that such a horse can carry a load of 300kg, more than its own body weight – compared with the Western cavalry of 1914–18 where the average horse carried half this amount.

Given the toughness of Mongol horses and riders, Genghis Khan could boast, 'It is easy to conquer the world from the back of a horse.' The *Secret History of the Mongols*, written during the reign of Genghis' son Ogedai, claims that Subedei chased his defeated enemies with a pole with a loop at the end, to catch them around the neck and break their spines or drag them for miles. It was not unusual for beaten rulers to be trampled to death by the Khan's horses.

Just as Father Carpini was appalled by what he saw of the Mongols, so Matthew Paris noted that their horses 'eat branches and even trees and [the Mongols] have to mount by the help of three steps on account of the shortness of their thighs'. Typical of Paris, who never saw a myth he did not record faithfully, this is nonsense; Mongols *jumped* into the saddle. It was impossible to keep a bad chronicler down. '[The Mongols] have misused their captives as they have their mares. For they are inhuman and beastly, rather monsters than men, thirsting for and drinking blood.' The khan's cavalry were reputed to slit a vein in a horse's neck and drink its blood when water was scarce on campaign.

The cavalry on the move must have been an impressive sight; thousands of horsemen trotting in silence, armed to the teeth. Special arrows were developed with a whistling whine (rather like the 'whizzbangs' of the First World War) which terrified enemy horses. Other arrows were specifically for bringing down horses and most lances had a hook at the end to rip horses' legs or unseat an opponent. The Mongol tactic of tying brushwood to their horses' tails to drag dust, giving the impression of vast numbers, was used in the desert regions of the Khwarazmian Empire (today's Iran and Iraq).

Genghis Khan crossed the Kizil Kum desert, which the Shah assumed was impossible, and destroyed Bokhara, the 'cupola of Islam'. He assembled the trembling imams in the central mosque and tore up the Koran with his own hands. 'I am the punishment of God,' he told them. 'If you had not committed great sins, he would not have sent a punishment like me.' He ordered the slaughter of 30,000 civilians over a two-day period. The mosque became a stable for his favourite horses.

By the time of Genghis Khan, the West had lost virtually all of its spiritual links with horses. Biblical references to horses come from the Old Testament, not the New, and Christianity had no room for animalistic myths. Among the Mongols, a stallion's soul lies in his mane, which is often kept after an animal's death. As with the Arab horses of the Muslims, a blue ribbon is tied around a sacred animal's neck. Such horses are never ridden, but may have been sacrificed at the deaths of great men. William of Rubrucque, travelling in Mongolia in 1253, saw a grave with poles extending above it, draped with horse skins and topped with horse skulls. Mares' milk was drunk at the funeral, just as Genghis Khan sprinkled some on the ground at his feet to ensure victory before a battle.

In the shamanistic traditions of the Mongols, a man's soul is the wind horse. In Tengenism, with its ninety-nine deities, Kisaya Tongri is a horse god who protects souls. At shamanistic rituals, drums are made of horse skin and traditionally, it is horses that carry their riders to the afterlife.

Genghis Khan died in August 1227 on campaign against the Tangut tribe. They were destroyed by the Mongol army overcome with grief and everything living in the Tangut capital was butchered. The perfect warrior was buried in the Burkhan Khaldun mountains. Forty slave girls and forty horses were buried with him; he would need them all in the afterlife. Then his cavalry, who had made him what he was, rode over his grave and it disappeared without trace.

Chapter 8

White Surrey

Count Rudolf of Lorraine sat his horse under his banner of the slanted cross. At 26, he was one of the most experienced soldiers in King Philip's army. He had fought in Lorraine itself, France, Brittany and Spain. He had pursued the Moors as far south as Gibraltar as an ally of Alfonso of Spain in the *Reconquista*. French chivalry had never changed since Senlac Hill back in the mists of southern England in 1066. Cavalry was the arm of God; the mounted charge the battle-winning tactic. Somewhere on the field around Rudolf the Valiant were bombards, primitive cannon and this was the first time they had been seen on a European battlefield. Rudolf found their presence vaguely amusing. If they did not blow up and kill their gunners, the damage they could do to the enemy was minimal.

Under him was the true weapon of any civilised war – the powerful destrier caparisoned with his personal arms, silver falcons on a red bar. Around him, his knights, armoured like himself in pig-nosed bascinets and padded jupons over their plate armour, watched the Genoese crossbowmen ahead of them forming up. Beyond that, the villages of Wadicourt and Crecy lay almost lost in the rainstorm that came out of nowhere. It was 26 August 1346 and high summer in northern France.

Between the French lines and the villages stood the English, their royal banner of the leopards and lilies in the centre. Rudolf had always resented that flag. Since he became the Duke of Lorraine at the age of 9, he had been taught that the English had no right to hold

French territory. Crecy itself belonged to Edward III, the arrogant king who was here today to defend it. Well, it was time to end this nonsense and drive the English out.

The crossbowmen advanced at the shouts, in an alien tongue, from their commanders. The rain was bouncing off their kettle-hats and turning the ground by the River Maye into a quagmire. France had few archers of her own, hence the importation of the Italians. The papacy had tried to ban crossbows in the twelfth century; they were too vicious for the knightly code of chivalry. But a crossbow bolt with its short shaft and wooden flights could pierce plate armour at forty paces and bring down horses. The Genoese would make short work of the English archers, unarmed and ragged with their feeble longbows that had to be destrung in wet weather because the hemp became warped.

Rudolf's cavalry gathered momentum, he, like the other commanders of the vanguard, trying to keep the pace steady, from the walk to the trot. Those damned crossbowmen were in the way, slowing things down. Then, a new downpour occurred, not rain this time but English arrows, a clothyard long, filling the air and thudding into the Italians who had no pavises to hide behind. Again and again the shafts flew, at three times the rate of fire of the crossbows and nearly twice the range. The Genoese broke, unable in the mud to press the strings of their bows to the ground, and Rudolf's knights rode them down, hacking about them with their swords.

Now. Now was the time to strike. Rudolf's sword was in the air and he roared his battle cry as the whole line thundered forward to destroy English claims to this land forever.

The battle of Crecy, fought in the early years of the Hundred Years War, ought to have been the end of the mounted charge and the reliance on heavy cavalry. In fact, because of hide-bound tradition and the snobbery of the French aristocracy in particular, the tactic survived for another 200 years and would be revived, usually with spectacular casualties, for many years after that.

As far as we can rely on the figures written down by Medieval chroniclers, the English losses were no more than three hundred. The French, however, suffered disastrously. Jean Froissart, the best-known contemporary historian, estimates 30,000 killed and wounded, but this has to be an exaggeration. Since the battle did not begin until late afternoon and continued into the twilight, some of the killing continued the next day, the English heralds wandering through the mass of bodies plastered in the mud, noting the heraldry on jupons and banners. Among the dead, they found the silver falcons of Rudolf of Lorraine with many of his knights around him. They found too the body of the blind king John of Bohemia with his three-feathered crest. He had had his horse tethered to his knights as he could not see to guide it himself; they all lay together in a heap.

What had won Crecy for the English was the longbow, actually a Welsh weapon made of yew. The speed, range and accuracy of the Welsh longbowmen impressed the English under Edward I during his conquest of Wales fifty years earlier and now Welshmen and, more importantly, their bows, were at the disposal of the king of England. Although bombards were fired at Crecy, it would be two centuries before artillery made its presence felt in battle.

The French cavalry charges – and there were several of them – exposed the weakness of the armoured knight. The archers aimed for the horses which were at once bigger targets than their riders, unarmoured and more likely to cause chaos. As the front line went down, those behind had to swerve or leap the obstacle of floundering men and terrified animals. Contrary to popular belief, a horse will not willingly tread on either man or horse and tales of destriers being trained to trample deliberately are nonsense. At Crecy – and it would be repeated at Poitiers in 1356 and Agincourt in 1415 – the French paid a heavy price for stubbornness and obsolescence.

> A good horse [said the fourteenth-century *Boke of St Albans*] should have fifteen properties and conditions, namely: three of a man, three of a woman; three of a fox, three of a hare and three of an ass. Like a man, he should be bold, proud and hardy; like a woman, he shall be fair-breasted, fair of hair and easy to lie upon; like a fox, he should have a fair tail, short ears and go with a good trot; like a hare, he should have a great eye, a dry head and run well; and like an ass he should have a big chin, a flat leg and a good hoof.

Allowing for the fourteenth-century slur on womanhood, which would leave today's feminists reeling, this was what experts looked for in any riding animal, but especially warhorses. By the time this was written, the average destrier cost £40, an exceptional one twice that. Most of the horses killed on Edward III's campaigns in the Low Countries between 1338 and 1340 were worth between £10 and £20. Only coursers, used for hunting, were this price. Palfreys, the riding horses depicted in the Ellesmere illustrations of Geoffrey Chaucer's *Canterbury Tales* (1386) could be bought at horse fairs such as the one every Friday at Smithfield, for between £4 and £5. Rouncies, hobbies, nags and hackneys were cheaper still and were used for haulage and short distances. Packhorses, for which special stone bridges were built over rivers, carried raw wool from farms and monasteries to weavers across England – they came in the 5 to 10 shillings range. A good saddle cost 5 shillings, halter, bridle and reins 6d to 1 shilling. The simple prick-spurs which riders had used by Chaucer's day for over 500 years were being replaced by the novel rowel type with a revolving spiked wheel. These cost 2 shillings.

It is a general historical fact that news and vital information throughout much of the past travelled at the speed of a horse. A palfrey could reach 40 miles an hour over short distances; a destrier far less. Most riders in the fourteenth century considered 15 miles a day a fair distance. Armies moving at the speed of their baggage trains could perhaps manage ten. Much of this depended on the availability of roads (the best in Europe were the old Roman military roads) and the time of year. Traditionally, the campaigning season was May to September, although it was rarely that neat. As we have seen, Hastings was fought on 14 October. Agincourt took place on 21 October and in very wet conditions totally unsuitable for cavalry charges. One of the fastest Medieval rides recorded was that of Edward I's messenger in July 1307 who rode 80 miles in a single day from Burgh on Sands (with a change of horses) to report the king's death to the Prince of Wales in London.

Expensive horses like destriers were given careful treatment by vets. In the winter of 1397, Henry of Lancaster bought a whole range of medicine for Lyard Gylder, a horse he had to leave behind in Calais on his return to England. They cost him 3s 7d. A second horse, Sorrell Blackwell, cost even more in vet bills – 5 shillings. In the case of damaged or poorly legs, vets recommended poultices of fat, honey and butter, not unlike the dressings given to wounded soldiers. The rest was common sense or leaving matters, as the Arabs did, to God's will; food, dry accommodation and rest were the best cures.

The importance of the horse in the later Middle Ages is borne out in that stealing one carried the death penalty. This was one of several principles that was carried over into the New World; a nineteenth-century cowboy without his horse could not actually be a cowboy at all! The fourteenth century was one in which animals could be put on trial, especially for killing their owners. In the case of horses, however, the punishment was slight. A horse that lashed out, quite possibly by accident and killed its master became the deodand – that is, it was given a value as a murder weapon – and became forfeit to the crown, since the king was the chief dispenser of justice in the country. By contrast, a vicious dog belonging to a poacher had a paw cut off!

Richard II's court in 1392–3 had eighty-nine valets of the stables and fifty-three grooms, but only two cleaners. Presumably, this pair had their work cut out.

As we have seen, the tournament was the training ground for squires who would fight as knights under, for example, Rudolf of Lorraine. The melee or behourd had become such a dangerous roughhouse that Henry II issued edicts against it as early as the 1170s and his son Richard the Lionheart granted licences to only five designated sites throughout England where jousts could take place. Such were the restrictions in England that several knights crossed the Channel to fight, especially when Henry III instituted fines, confiscation of lands and disinheritance of children in the 1260s. The Church too railed against this kind of competitive sport, issuing five papal bulls in less than a century. Nothing worked; as historian Ian Mortimer notes in *The Time Traveler's Guide to Medieval England*:

> Where else, in all history, can you see the richest, most powerful and most privileged members of society risk injury and death for the sake of your entertainment? Where else, in all history, can you find rich and powerful men *paying* for the privilege of breaking their necks and goring each other in public?

By the time of Crecy, the tilt, fought between individual knights in the lists, had largely replaced the melee. Castles had flat areas called the tilt for this purpose – Kenilworth in Warwickshire has one, so has Carisbrooke in the Isle of Wight. The rules of these mock battles were rigid, laid down by marshals at arms who oversaw the proceedings. Each pass was called a run, two knights facing each other on horseback with a shield, strapped to their left arms for defence, and a lance, usually made of ash, twelve feet long. The idea was to hit the opponent on his shield or helmet and knock him out of the saddle. Three lances were usually broken in those runs as the knights collected new weapons at the end of the tilt yard from their attendant squires. Attacks were delivered at the gallop, so that the impact of the destriers weighing up to 1,000lb each and their 200lb riders was colossal. Half a ton slamming into shield or helmet was almost bound to cause damage; still more when an unhorsed competitor hit the ground. Occasionally these duels were fought to the death or were legal combats over land ownership or affairs of honour. We know that Sir John de Creke of Westley Waterless in Cambridge, whose memorial brass tomb is in the church there, fought a land duel (which he won) in the 1320s. In 'friendlies', blunted weapons, called the lance of peace were used. The lance of war had an iron tip. In rare cases, these deadly points could penetrate a knight's vizor; in 1559, Henri II of France was killed in this way, the lance point slicing through his brain. This accident led directly, in the king's absence, to the French Wars of Religion. Perhaps in the same decade, the Hungarian nobleman Gregor Baci had his skull pierced with a lance. The tip entered through his right eye socket and the only thing the surgeons could do was to saw off the lance. There is a bizarre portrait of the man in an exhibition of freaks in Ambras Castle, Innsbruck, Austria. Medical experts today believe that if the lance missed most of the brain, Baci could have lived a year, perhaps more, with the weapon still embedded in his skull. Ironically, the lead point on the lance shaft would have acted as an antiseptic.

Women's roles in Medieval tournaments were strictly those of spectators. In keeping with the romances and the hugely popular concept of courtly love from the twelfth century, they gave their favours (coloured scarves) to their husbands/lovers to wear in tournaments. Several events in England were held in honour of the legendary King Arthur, the most spectacular perhaps at Caernarvon, Wales, in 1284.

The English horses connected with royalty are known to us only through Shakespeare. In Act V, scene V of *Richard II*, the newly deposed king, facing his murderers at Pomfret (Pontefract) castle, is gloomily told by a groom how the usurper, Henry Bolingbroke, has taken everything from the ex-king, even his horse:

> O, how it yearned my heart when I beheld
> In London streets, that coronation day,
> When Bolingbroke rode on roan Barbary
> That horse that thou so often has bestrid,
> That horse that I so carefully have dress'd.

Richard asks –

> Rode he on Barbary? Tell me, gentle friend,
> How went he under him?

The groom is as embarrassed as Richard –

> So proud as if he disdain'd the ground.

To which Richard adds –

> So proud that Bolingbroke was on his back!
> That jade hath eat bread from my royal hand;
> This hand hath made him proud with clapping him.
> Would he not stumble? Would he not fall down,
> Since pride must have a fall, and break the neck
> Of that proud man that did usurp his back?
> Forgiveness, horse! Why do I rail on thee,
> Since thou, created to be awed by man,
> Wast born to bear? I was not made a horse;
> And yet I bear the burthen like an ass,
> Spurr'd, gall'd and tried by jouncing Bolingbroke.

Shakespeare's source material for his history plays was usually Ralph Holinshed and Edward Hall, chroniclers who were little more than Tudor hacks, employed to extol the virtue of the house that usurped the English throne in 1485. Whether Roan Barbary was *actually* Richard II's horse is therefore debatable. The colour speaks for itself and the Barbs were a breed similar to the Arabs, but less spirited. Intelligent, able to function on meagre rations, gentle and sure-footed, such horses were favoured by the Berbers of the Maghreb, hence the name of the breed. By the sixteenth century, the whole of the North African coast was called Barbary.

The other royal horse to come from Shakespeare is White Surrey, ridden by Richard III at the Battle of Bosworth on 22 August 1485. If Shakespeare's Richard II is ambivalent – he is a bad king, but we feel for him by the time of his murder – Richard III is a monster, too well-written to be a pantomime villain. As the man overthrown by the Tudors, Richard was bound to be blackened out of all recognition by a playwright who craved patronage and knew on which side his bread was buttered. On the evening before Bosworth, Richard is full of foreboding. According to Shakespeare, he was guilty of eleven murders, including those of his wife and nephews, so that is hardly surprising. Full of nervous energy, he fires orders to his lackeys. To Richard Ratcliffe, he says –

> Fill me a bowl of wine. Give me a watch [guard].
> Saddle White Surrey for the field tomorrow.
> Look that my staves be sound and not too heavy.

As with Richard II's Barbary, we only have Surrey's name from Shakespeare and it may have been a corruption of Syrie (Syria), implying Arab blood. The staves that the king refers to are lance shafts, traditionally made of ash, which we know he used at Bosworth. While Barbary is simply a royal horse of which no details are known, Surrey played a major – and fatal – part in the last battle to witness the death of a king in England.

While the vanguards of the royal army struggled with the Earl of Oxford's on the slopes of Ambien Hill in the Leicestershire countryside near Sutton Cheney, Richard swept around them on Surrey, leading a cavalry charge to kill the would-be usurper Henry Tudor and consign him to a footnote in history. Just as there was a marsh on the field at Senlac and the watery ground at Bannockburn, so a quagmire at the foot of Ambien Hill slowed Richard's charge. He skewered William Brandon, Henry Tudor's standard bearer, with his lance and catapulted the huge John Cheyne out of his saddle. Then Surrey got stuck in the cloying mud and Richard's enemies closed in. We know from the bones found almost miraculously under a car park in Leicester which are indisputably those of the king, that

Richard was killed by a series of blows to the head. Surrey's fate is unknown. One modern account has him dying with his master, but there is no contemporary mention. Later stories have the dead king's body slung over the back of a horse that is clearly not Surrey, but that, if true, does not explain the animal's fate.

Richard III has only recently been re-evaluated as a good king and a balanced individual rather than a crook-backed psychopath. It was in the seventeenth century that a kinder view of the last Plantagenet emerged for the first time. In the English Civil War, General 'Black Tom' Fairfax, commanding Parliament's New Model Army, hailed from Yorkshire, where Richard, in his day, had been extraordinarily popular. Fairfax called his horse White Surrey in honour of him.

Chapter 9

Tziminchak, the Thunder and the Lightning

Father Juan de Orbita was a Franciscan and he had one goal in life – to convert the poor misguided Itza people to the one true God worshipped by the Christians. It was 1618 and Orbita and his fellow-missionary, Bartolome de Fuensalida, had spent six months travelling by sea, river and land to the Itza capital at Tayasal. The Franciscans had been welcomed, given food and water and had a great fuss made of them. In the weeks that followed, with a mixture of sign language and the help of interpreter guides, de Orbita had explained the true religion to the Itza chief, Canek. The man had listened politely and even asked questions, but it was impossible for his people to give up their gods for somebody else's. When Orbita asked why, Canek offered to show him.

Neither missionary was ready for what they saw. A great procession, with feathers and robes, made its way through impenetrable jungle, Canek and the priests at its head. They came to a clearing and there stood a huge temple, built in the pyramid style of the Mayans. The guards swung open the great gates and the priests went inside. This was a cathedral, they realised, like those they knew at home, but there were no crosses, no swinging incense, no crucifix or statue of the Virgin.

Instead, in the centre where the high altar would have stood in Spain, was a statue of stone, painted black, with flaming scarlet nostrils and mouth. It was a horse, sitting on its haunches and around its hoofs were scattered the loving offerings of the people

– flowers and fruit, like the garlands on saints' days in Madrid or Burgos. This, Canek's interpreters explained, was their chief god, Tziminchak, Lord of the Thunder and the Lightning. And he was a horse.

The instinct of the Franciscans was to smash the thing as Moses did in the Old Testament. The worship of idols was unthinkable to these men, even though they had done exactly the same all their lives, praying to artificial images of a man who had been crucified 1,600 years earlier. At least Tziminchak was a new deity, part of the local Itza tradition in a way totally different from Jesus Christ, who had no links with Medieval Spain at all.

Tziminchak had originally been a real, flesh and blood animal, Morzillo (black) the warhorse of the conquistador Hernan Cortes who brought horses to the New World and helped to create a new way of life.

There is a deep irony in the fact that North America, where the oldest horse remains were found, had none left by the sixteenth century. Eohippus must have been hunted to extinction or wiped out by disease as there is no record of any kind until the conquistadors (conquerors) arrived in search of God, gold and glory. In the late fifteenth century, with the Ottoman Turks riding their Arabs attacking merchants' caravans to the Middle East, there was a need to find a new route for the silks and spices that were the luxury perks of the seriously rich throughout Europe.

Two of the most enterprising rulers in pursuit of a fast buck were Ferdinand and Isabella of Spain. An edict from them, dated 23 May 1493, reads:

> We command that a certain fleet be prepared to send to the islands and mainland which have newly been discovered ...

and among those we command to go in the vessels there shall be twenty lancers with horses ... and five of these shall take two horses each and the two horses which they take shall be mares.

The previous year, Christoforo Colon (Columbus) had sailed west to reach the east in a bizarre belief that the world was round. He had sailed in Spanish ships with Spanish crews and rammed the flag of Castile and Aragon into the sand of the beach where the *Santa Maria* berthed and called the island San Salvador – the saviour. To his dying day, Columbus believed that he had found India, the Spice Islands and the name stuck for the locals he found there for five hundred years. It is unknown whether he realised that there were no horses in the West Indies or the American mainland, but clearly, Ferdinand and Isabella intended to provide their own Spanish bloodstock, almost certainly Andalusians. But the wranglers in charge of the animals sold them and substituted 'common nags' instead.

That glitch aside, there was a stud farm in Hispaniola by 1500 and another in Cuba after its conquest fifteen years later. It was on this island, according to legend, that the Spanish warhorse became a god. During the brief campaign for Cuba, the hidalgo Pánfilo de Narvaez found himself facing a night attack from several hundred tribesmen. Still in his nightshirt, he leapt on to the back of the only horse with his company of archers and galloped into the attackers. The locals had never seen a horse before and assumed that man and animal were one, a centaur-like apparition which terrified them. They ran and depopulated their villages in terror.

Four years later, one of Narvaez's company, Hernan Cortes, left Cuba with 11 ships, 800 soldiers, a collection of African slaves and local women and 16 horses. They landed on the Mexican coast on 12 March and the rest was history. With the expedition was

Bernal Diaz del Castillo who wrote a very detailed account of the conquest fifty years later with something approaching total recall. Astonishingly, we have a complete record of the first 'new generation' of horses in the New World and their owners. Cortes himself rode a chestnut stallion which died at San Juan Ulna. Undeterred, the commander helped himself to a black called El Arriero (Drover) that he renamed Morzillo, the horse that would become a god. Several horses had joint ownership, probably because of plans to breed them. Pedro de Alvarado and Hernando Lopez de Avila owned a bay mare, excellent for racing and tilting. Alonzo Puertocarrero was probably not much of a horseman – he was happy to swap his grey mare with Cortes for a gold shoulder knot. Juan Vesasquez de Leon rode another grey, La Rabona (Bobtail). Cristobal de Olid had a brown horse, unremarkable, but, like most of them, a hardy campaigner. Francisco de Morla's bay had a good mouth and was very fast. Perhaps Francisco de Montego's and Alonzo de Avila's chestnut was too skittish, because Diaz noted that 'he was no good for war'. Likewise, the bay with three white stockings ridden by Juan de Escalante 'was not very good'. Diego de Ordas had a barren grey mare, not very fast, but Gonzala Dominquez's roan was excellent. Pedro Gonzalez de Trajillo's bay 'galloped well' and the unfortunately named Moron was a piebald ('pinto') with white forelegs. The other pinto, however, belonging to Baena of Trinidad 'proved worthless'. Lares' bay was impressive at the gallop and the last horse to be described, the brown mare belonging to Juan Sedeño gave birth on board ship. Sedeño, Diaz noted with a certain waspishness, was the richest man in the fleet, with his own ship, a black slave and 'much cassava bread and bacon'.

The first herd included eleven stallions and five mares and Cortes probably hoped for more. None was available, however, so that his cavalry force was peculiarly under-strength in comparison with his

500 infantrymen. His horsemen were ginetes, lightly armed with Moroccan leather shields, swords and javelins.

As in Cuba, the opposition, outnumbering the Spaniards perhaps 300 to one, fled terrified and the horsemen rode among them, skewering them with their lances. The problem with the conquest of Mexico, despite the meticulous detail of Diaz, is that it lends itself to legend and tall tales. Cortes is said to have invited friendly locals to see his stallion strutting his stuff in front of a mare who was in season. The ground-pawing, eye-rolling and head-tossing which is part of the mating ritual horrified the watchers, as Cortes knew it would. He led the stallion away and explained that he had ordered the horse not to hurt them because of their friendliness.

The reports reaching the Aztec king Montezuma at Tenochtitlan were terrifying. White-skinned men with beards had reached the coast on floating islands and they brought monsters with them:

> These stags [the Aztecs did not know what else to call them] snort and bellow. They sweat very much ... the foam from their muzzles drips onto the ground. It spills out in fat drops like the sap of soap-plants ... They make a great din, as if stones were raining on the earth. The ground is pitted and scarred where they set down their hoofs ...'

It would be two years before Montezuma saw these beasts for himself, with the conquistadors battling Aztec tribes, steeling themselves against what they thought were invincible odds – the deadly combination of horseflesh and firepower, as well as disease and difficult terrain. For days, there was no water and when there was, the stagnant jungle pools were undrinkable. In battle, the Aztec archers were formidable – their shafts were longer than crossbow bolts and the Spanish guns, for all the terror they brought to the battlefield, were not always reliable.

Inevitably, the locals came to realise that the monsters were actually mortal. When a mare went down to an Aztec axe, her body was dragged away and eaten. 'They killed a horse,' Cortes wrote in one of his regular reports to the emperor, Charles V, 'and God alone knows how great was its value to us and what pain we suffered at its death, because, after God, our only security was the horses.' He may have been paraphrasing the Biblical Book of Proverbs – 'The horse is prepared against the day of battle; but safety is of the Lord.' There were times in that grim campaign, however, when the Spaniards ate their horses too.

Tenochtitlan fell after a gruelling ninety-day siege in August 1521. After it, Cortes led another expedition, marching nearly 1,000 miles south to what is now Honduras in Central America. The balance of his cavalry and infantry was better this time – 120 horsemen and perhaps 200 footsoldiers. He also had nearly 2,000 Aztecs, happy to throw in their lot with the invaders. This would become a pattern in the centuries ahead; when the British and French fought for American territory in the Seven Years War and the colonists then fought the British, both sides used 'Indians' as trackers and guides.

Once again, Cortes used his horses as weapons of terror, much as, in the ancient world, Hannibal of Carthage had used elephants against the Roman legions. The campaign was devastating. The Spaniards had only forty men still in their saddles, the Aztecs perhaps fifty. None of the horses was fit enough to continue.

At Tayasal, within four days' ride of the Honduras coast, Cortes' horse Morzillo had to be left behind. He had carried various riders in the New World for nearly fifteen years and had been wounded several times. Cortes wrote to Charles V, 'My Black got a splinter in his foot and was unable to go on. The chief promised to cure it, but I do not know what he will do with him.'

What the Mayans did was to make offerings to the horse in the belief that he was supernatural; the diet, ironically, probably killed him. Terrified that Cortes would return and blame them for the animal's death, they erected a mausoleum over Morzillo's grave and erected the statue that the Franciscan missionaries were shown nearly a century later. Because the conquistadors' horses made the ground shake like thunder and because their guns flashed like lightning, the new god was called Tziminchak, the thunder and the lightning.

'On clear moonlit nights,' today's inhabitants of the town of Remedino will tell you, 'you may see Tziminchak deep in the waters of the lake, patiently receiving the worship of the Itzas while he awaits Cortes' return.'

Chapter 10

Black Barbarie

The furious German comes, with his clarions and his drums,
His bravoes of Alsatia and pages of Whitehall;
They are bursting on our flanks! Grasp your pikes! Close your ranks!
For Rupert never comes, but to conquer or to fall.

The Battle of Naseby
Lord Macaulay

Prince Rupert of the Rhine sat on Black Barbarie at the bottom of the slope. He had fewer than 2,000 horsemen behind his standard of the Palatine, seasoned troopers who had come through a dozen battles since Powick Bridge, led by gentlemen born to the saddle. This was going to be a good day. The prince would have liked his hunting dog, Boye, to be with him, but Boye was dead, killed at Marston Moor nearly a year ago. Men said the animal was a familiar, an imp serving the devil, and that Rupert was a witch. If the Scots caught him, they would burn him; if Parliament's English did, it would be the rope. Fanning out to his left the standards of Jacob Astley's infantry shifted in the breeze, the sun glinting on their kettle-hats and catching the deadly points of their eighteen-foot pikes. Beyond them, although they were out of Rupert's line of vision, the left wing of the Royalist cavalry, Marmaduke Langdale's Northern Horse, waited for the word of command.

Rupert was the most experienced officer in the army of his uncle, King Charles I. But Charles was stubborn, with his haughty manner, his set ways, his irritating stammer and his ghastly wife. Too often, it seemed to Rupert, when he had argued with the feather-bed officers who fought for the king, Charles had supported them, listened to their whingeing, sympathising with their lot. Before Edgehill, in October 1642, not one of them had fired a shot in anger. Rupert had. He had been fighting in the saddle since he was 15 and had captured Breda a year later. During his three years in prison in Linz he had devoured every book on tactics he could get his hands on. These English cavaliers did not know a tactic from a hole in the ground. But they were, at least, natural horsemen and he was glad of that.

To Rupert's left that day at Naseby, he could see the Parliament's dragoons forming up. Typical, the cowardly bastards were hiding behind a hedge, their horses at the rear, tired old nags that could barely stagger to the field. The men were fumbling with their matchlocks, poking the barrels through the tangle of thickets. Was that Colonel Okey's standard flapping above them? Rupert was not sure.

What he was sure of was the plodding riffraff facing him across Broad Moor to the south. And that *was* a flag he knew – Henry Ireton. The man was a lawyer from Nottinghamshire; no reputation and little experience. Rupert remembered him from Marston Moor, but Marston Moor had been a disaster – Rupert had not lost the battle and Ireton had not won it. And behind the lawyer, Rupert could see the cavalry of … what did they call it? The New Model Army? Well, it may have been new, but he doubted it was much of a model. He was not sure it was even an army.

Scheiss! Rupert was furious, pounding Barbarie's saddle as he realised the situation. Old Ironsides was there, as Rupert knew he would be, but the old hypocrite was way across the field, commanding the Parliament's left wing. Well, no matter; there would be time to deal with Noll Cromwell later. Deal with Ireton first.

Rupert drew his sword, kissing the steel curves of the swept hilt and urged Barbarie forward. The trumpet sounded behind him and the whole line lurched to the walk, horses' heads tossing, mouths champing at the bit. Each man stood upright in his saddle, his sword erect in his hand. Most of them, like Rupert, had fastened the straps of their plumed pots under their chins. They scorned the barred lobster-tails of the enemy, but going into battle bareheaded made no sense at all.

Over the past two years, Rupert had learned a great deal. In particular, he needed to be cautious. 'Black Tom' Fairfax, commanding the army ahead, had twice the king's numbers and timing would be everything. He kept Barbarie in check, feeling him pull against the bridoon rein, arching his neck and kicking his heels high. Astley's Foot were moving too and Rupert had to keep pace with them. He could see the massive block of Parliamentary infantry moving, Fairfax's standard in the centre of the van.

Shots rang out in the morning, puffs of smoke from Okey's dragoons. Those men had been on patrol all night; they were tired and off their stroke. Only one of Rupert's men went down, another wheeled his horse away as the animal staggered and fell. They were barely within range and if Ireton came on too fast and too soon, they would be shooting their own men.

Then there was more firing from the front. The Forlorn Hope of both sides had clashed in the centre of Broad Moor. A few scattered musket shots, shouts and obscenities and both units broke for the safety of their own lines. Rupert could hear a thunderous noise from Fairfax's Foot; the bastards were singing a psalm. He could even see their chaplains, sitting ducks in their black and white robes. Whoever God gave the victory to today, it would be via men with steel in their hands and steel in their hearts – those damned Puritans had nothing to do with it.

Now, Rupert judged the time to be right. His men would be riding uphill, it was true. All the more need, then, for the gallop. He held his sword high and rammed his spurs into Barbarie's flanks. The stallion jerked forward, free at last from the tight rein and thudded forward at a canter, the others close behind. Okey had all but stopped firing now – Ireton's cavalry were moving forward, using the momentum of the slope to close with Rupert's troops.

'God our strength!' Ireton's troopers yelled, hacking and slashing at Rupert's horsemen, but either the Lord was not listening or he had pitched his tent with the Royalists that day. Rupert hacked an officer out of the saddle, ducking under a flurry of blades. He slashed the arm of Ireton's flag-bearer, ripping it from wrist to shoulder, but the lawyer himself, as slippery in the saddle as he probably was in court, wheeled away out of Rupert's reach and lived to fight on.

His cavalry did not hold for long. Rupert noticed that they were riding close together, knee to ham but they were only moving at the trot, trying to fire their wheel-locks as they came on. Rupert's horsemen galloped through them, knocking them aside, jerking them out of the saddle, riding over them. They rode on blindly, slashing and stabbing as Ireton's Horse reeled backwards, struck by the thunderbolt that was the Royalist Horse.

There was their prize. There – ahead. The Parliament's baggage train. There would be guns, ammunition, spare horses, perhaps Old Noll's Bible unless he had that stuffed up his arse. *This* was war. *This* was what Rupert could do. And the field of Naseby, in the year of Cromwell's Lord 1645, was living testament to that.

Ninety years ago, Sellar and Yeatman wrote their incomparable send-up of school history textbooks called *1066 And All That*. It was gloriously silly and has not been bettered since, not even by the *Horrible Histories* series. According to Sellar and Yeatman, the

Civil War can be summed up as the Royalists being 'wrong but romantic' and Parliament 'right but repulsive'. They use the generic term 'Roundhead', which even serious historians used at the time. The Roundheads were actually London apprentice boys who wore their hair cropped short and could be relied upon to provide a rent-a-mob at every opportunity. Since London was almost universally behind Parliament, the Roundheads became associated with them. As to the men who led Parliament's armies in the Civil War, people like Fairfax, Cromwell and Ireton, they wore their hair as long as the cavaliers they were fighting. Cavaliers is a *slightly* more accurate term than Roundhead in that the king's horse officers were almost entirely aristocracy and gentry. Cromwell famously sneered at them, 'I would rather have a plain, russet-coated captain who knows what he fights for and loves what he knows than that which you call "gentleman" and is no other.'

The Civil War was a power-struggle between Charles I on the one hand, claiming divine authority to rule as he saw fit, and Parliament, anxious to establish its powers under an oligarchy. How 'right' they were, in Sellar and Yeatman's terminology, is open to question. 'Right' in the seventeenth century meant tradition and law – and in that respect, right lay with the king. Parliament, by comparison, was a relatively new invention, originally a body of advisers to the king. The Speaker, who still chairs debates in the Commons, was a royal servant who reported to the king what was discussed in the Westminster chamber. During any periods of weakness of the monarchy, Parliament pushed the boundaries of what was always a delicate balancing act.

Elizabeth was a strong monarch, but she was a woman – the only one in government – and that did not sit all that well with the pushy gentlemen who represented the shires. Her successor, James VI of Scotland and I of England, had nothing like the negotiating skills of the queen. He hated witchcraft and tobacco, was almost certainly bisexual and rode roughshod over his parliaments. His son, Charles, did not even possess the limited finesse of his father and by 1629 had quarrelled with Parliament to the extent that he tried to rule without them.

Strapped for cash in a semi-religious war he had launched against the Scots, Charles was forced to recall Parliament in 1640, but the 'eleven years' tyranny' must be punished and relations remained strained for two years until Charles tried to arrest five ringleaders who opposed him and left London to raise his standard in Nottingham. The unthinkable had happened; England was at war with itself.

The last battle fought on English soil, other than clashes with the marauding Scots, had been at Stoke, the final conflict in the Wars of the Roses in 1487. The Tudors had no standing army and had systematically broken the power of their over-mighty subjects by taking away their private armies. English military involvement in Europe in the sixteenth and early seventeenth century was limited and small-scale and did little for the reputation of British arms.

When war broke out in 1642, both sides used the Commissions of Array to raise troops and frantically organised militia in their areas of influence. Of these, the London Trained Bands were far and away the most professional; the others were frankly amateur. As Cromwell wrote to John Hampden after Parliament was defeated at Edgehill, Warwickshire in October 1642, 'Your troopers are most of them old decayed serving men and tapsters and such kind of fellows ... You must get men of a spirit that is likely to go as far as gentlemen will go, or else I am sure you will be beaten still.'

The problem with many men's comments on the warfare of the 1640s is that it was heavily influenced by Puritan sentiment which today would be classed as religious mania. Cromwell had organised a superb cavalry force by 1645 – the New Model – but he still had this hypocritical approach to it all. Of Naseby, in June 1645, he wrote, 'when I saw the enemy draw up and march in gallant order towards us, and we a company of poor ignorant men …' he neglects to tell us that his 'poor ignorant men' outnumbered their enemies two to one.

If men in England had little experience of recent warfare, the opposite was true of Europe. The Thirty Years War (1618–48) raged across what today is Germany and dragged in most European nations, including Sweden, whose king, Gustavus Adolphus, was one of the great innovators of the battlefield. The day of the heavily armoured mounted knight had gone and 'to trail the pike' was now the honourable skill in battle. Pikemen, with their armour and their eighteen-foot pikes, had a special tactic for defending against cavalry. A pikeman would ram the butt of his weapon against his instep with his leg out straight behind him and the pike-point at an angle to skewer a horseman or stab a horse in the eyes. These formations were called *tercios* in Spain, but it was Adolphus' Swedes who became the experts, taking over from the Swiss and German *landsknechte* of the sixteenth century.

Cavalry tactics were badly affected in this period by the increasing use of firepower. The early matchlocks and wheel-locks were slow to load, clumsy and inaccurate, with a short range. For a man to load, fire and hit his target one-handed from the saddle was asking a lot and few were good at it. As these firearms were replaced by flintlocks by the end of the century, the problem became more acute. A musketeer no longer needed the protection of pikemen, because he could reload and fire at speed.

Adolphus' answer was for cavalry to advance at a trot. Cromwell, who adopted the tactic, calls it 'a pretty round trot'. Only the front line carried firearms and they fired together in a volley. They did not reload but threw their pistols and carbines (short muskets) at the enemy before drawing their swords and hacking at the opposition. This was not *that* successful, as many good and expensive weapons were lost that way, but it proved surprisingly effective on a number of battlefields. Adolphus was killed at Lutzen in 1632 but his ideas lived on. We know that Cromwell, who had no military experience before Edgehill, owned two books on the Swedish drill that Adolphus had formulated.

Undoubtedly the most experienced soldier that Charles I had was his nephew, Rupert of the Rhine. The problem was that Englishmen did not like taking orders from a foreigner, especially when that foreigner was trying to introduce techniques and tactics they had never heard of. Rupert was by nature a hothead and Adolphus' trotting cavalry held little interest for him. In this, he and his English cavaliers – his 'bravos of Alsatia and pages of Whitehall' – agreed. The cavalry charge must be delivered at the gallop, with the maximum penetrative force – and the use of the *arme blanche*, the sword. In two crucial battles – Marston Moor and Naseby – Rupert's headlong charges scattered the enemy, but then he was not able to rein them in and both actions ended in defeat.

The heaviest cavalry dominating the battlefields of the Thirty Years' War and the Civil War were the cuirassiers. Technically, the cuirass was a breast plate and it had been in use since ancient times. The seventeenth-century version was made of reinforced steel and 'tested' by having a pistol ball fired at it. The ball would dent the metal but not penetrate, so it was the forerunner of modern bullet-proof vests. Cuirassiers, however, were armoured from head to knee, with thick rawhide boots below that. Because the arms and thighs were covered in hinged plates to give flexibility, they were known as 'lobsters'. The only unit in the English Civil War equipped like this was the regiment of horse raised by Arthur Hazelrigg for the Parliament. The heaviest armour made (90lb) is on display in the Landeszeughaus museum in Graz. Its bulk explains Edmund Low's complaint after Edgehill – '... neither could I find my servant who had my cloak, so having nothing to keep me warm but a suite of iron, I was obliged to walk about all night, which proved very cold by reason of a sharp frost.' On the other hand, cuirassier armour saved lives. In a duel at Roundway Down, 13 July 1643, Captain Richard Atkyns clashed with Hazelrigg. The royalist 'tried him [attacked him] from the saddle and could not penetrate him nor do him any hurt'. Atkyns in the meantime had his sword arm crippled and his horse's nose slashed by the cuirassier's sword.

Cuirassier armour was the accepted 'full dress' of everybody from Charles I in the 1630s to John Churchill, Duke of Marlborough, in the early years of the eighteenth century, although neither man ever wore these defences in battle. On the Continent, cuirassiers abandoned lances and used wheel-lock pistols, a brace carried in holsters on both sides of the horse's neck.

Medium cavalry wore breast and back plates over thick hide coats (the russet coats that Cromwell refers to in the quotation above). Such coats were very expensive and very few of them have survived. The large, heavy boots could be worn up to knee level or rolled down and the long, straight sword (some with 'mortuary' hilts after the execution of Charles I) was the most common weapon, although a poleaxe was sometimes carried too. The 'pot' (helmet) had a lobster-hinged neck plate and a raisable visor with three bars to protect the face. These men, of the New Model Army raised by Cromwell and Fairfax, came to be called Ironsides, not because of their armour but after Cromwell himself to whom Prince Rupert paid the grudging compliment.

In the New Model, men received 2 shillings a day, but, as in all wars, pay was often delayed. A cavalryman had to find his own food and the forage for his horse; even his horseshoes and lodgings came out of his own pocket. Officers, of course, fared better; a lieutenant was paid 5s 4d a day, a captain 10s, a major 15s 8d and a colonel 22s. The New Model had 7,000 cavalry, divided into eleven regiments. Each one had six troops of 100 men.

It is difficult to be accurate about the sort of men who joined the Ironsides. On the one hand, they were 'brutal, licentious soldiery' in common with the average fighting man throughout history. On the other, many of them were devout Puritans, with a rabid hatred of popery and royalty and Cromwell's chaplains made sure that they were drip-fed such propaganda consistently. Most of them were from London, Kent and East Anglia, with a preponderance of urban recruitment. The Royalists, for their part, drew most of their support from the north, the west and the rural areas. The bottom line of this was that the king's army was usually outnumbered and always strapped for cash. When Lord Macaulay wrote *The Battle of Naseby* poem in 1854, he attributed it to Obadiah Bind-Their-Kings-in-Chains-and-Their-Nobles-With-Links-of-Iron, sergeant in Ireton's regiment. This was clearly tongue-in-cheek, but Puritan names genuinely included Fortitude, Grace and Killsin, implying the incomprehensible religiosity in the Parliamentary forces. The real 'madmen' had already sailed west on board the *Mayflower*. Never before or since have cavalrymen been such political and religious figures.

The horses of the Civil War have defeated some historians. Brigadier Peter Young, one of the founders of the first British re-enactment groups, the Sealed Knot, confessed, 'it is not easy to explain what they were'. And the best he can come up with is that they were 'more like heavy hunters or the modern chargers of the Household Cavalry'. Horses had been imported to England by royalty throughout the Middle Ages; we have encountered the Spanish mounts of William the Conqueror, the Barb of Richard II and the possible Arab-blood breed of Richard III. Henry VIII made his famous joke about the appearance of wife-to-be Anne of Cleves as his Flanders mare because he imported horses from northern France and the Low Countries regularly.

The warhorse of the seventeenth century was a native-bred variant of the great horse, which explains Peter Young's comparison above. They would have looked like the drum horses still used by today's British cavalry for parades – 16 hands high with a powerful build and heavy feet. Such animals were trained in the menage, a variety of movements from the caracole (effectively running on the spot) to the gallop. Rather as important men were depicted in cuirassier armour, so they were shown riding great horses; Van Dyke's paintings of Charles I are typical. The animals have impossibly solid necks, shoulders and rumps and their manes and tails are worn crimped and long. For one of his paintings, the artist paid sixty 'pieces' (of silver) as opposed to the twenty pieces that a cob would cost. Dragoons rode cobs; so did women and children.

Gervase Markham sang the praises of English breeds in 1617:

> Some former writers ... have concluded that the English horse is a great jade, deep-ribbed, sid-bellied, with strong legs and good hoofs, yet fitter for the cart than either saddle or working employment. How false this is all English horsemen know ... The true English horse ... is of tall stature and large proportions; his head though not so fine as either the Barbarie's or the Turk's, yet is lean, long and well-fashioned; his crest is hie ... his chyne be straight and broad and all his limbs large, leane ... and excellently jointed.'

In the 1650s, when the dust of war had settled, the Duke of Newcastle became the arbiter of all things to do with equitation, and many of the engravings and the descriptions of horse training come from that period, showing horses being put through their paces as in today's dressage competitions. One particular trick, not practised today, is the double kick with the back legs, designed to defend a cavalrymen from attack from behind.

The quality of a mount affected how much Parliament was prepared to pay for its use. Man and horse rated 2s 6d a day, a horse only half that. The value of the horse was assessed and it was lent to Parliament with a promised repayment of 8 per cent interest. This became a forced loan, as more cavalry were needed by the New Model. This, in the form of Charles I's tax called Ship Money, had been one of the flashpoints that led to war in the first place and here was Parliament blatantly using the same tactic. All Parliament's horses were branded on their hindquarters. By royal proclamation in 1642, the king requisitioned 'horses, geldings, mares or nags be sent in … to be used as Dragoon horses and with them their saddles and bridles.'

Horse armour had disappeared by this time as slowing up animals both on the march and in the field. The saddles of the period do not look very different from Medieval types, padded with rolls and with a high pommel and cantle. Stirrups were still worn long.

The horses of the New Model cost on average £7 10s. Dragoon cobs were £4 and draught horses for artillery and baggage, £6. The artillery of the period was increasing in numbers and variety but most gunners were happiest when pounding away with their heaviest shot against static fortifications. Although mention is made occasionally of a 'galloping gun', there were nothing like the horse artillery of later generations.

Chapter 11

A Gallop Through the Age of Reason

'Barbarism,' wrote the French encyclopaedist Jean le Rond d'Alembert, 'lasts for centuries; it seems that it is our natural element. Reason and good taste are only passing.' In terms of philosophy, the Renaissance of the sixteenth century, beginning in Italy as an attempted rebirth of classical Roman and Greek cultures, morphed into the Scientific Revolution of late seventeenth-century England. From there emerged the period of Enlightenment, Voltaire's Age of Reason. It is a sad indictment on the human condition, however, that while a handful of men (and virtually no women) questioned deep philosophical issues, most people just got on with the job of living and dying. And they went on killing each other too.

Frederick II of Brandenburg-Prussia was a philosopher, but he was also a king and a brilliant soldier. His ex-friend, the arch-cynic philosopher Voltaire, wrote:

> The most determined flatterer will readily agree that war always drags plague and famine in its train if he has glimpsed the hospitals of the armies in Germany and has been in certain villages in which some great warlike exploit has been performed. It is certainly a very fine art that desolates the countryside, destroys dwellings and brings death to 40,000 [to] 100,000 men in an average year.

When Napoleon visited Frederick's tomb after his victory over the Prussians at Jena in 1806, he said to his marshals, 'Uncover, gentlemen. If he were alive, we would not be here.'

In England, the Restoration of the monarchy saw a renewed interest in horse breeding. Charles II, at 6ft 2in, hardly had the build of a jockey, yet he unaccountably won every race he entered at Newmarket! He sent his Master of Horse to North Africa to buy good quality stallions and brood mares. Although Charles' principal interest lay in racing, the cavalry horse required the same strengths – he must be fast, with great stamina *and* be able to jump the five-barred gates that, with the coming of eighteenth-century enclosures, were becoming much more common. Because of the importation of Arab stock, characterised by the Byerly Turk and the Darley Arabian, famous sires on whom no price could be put, the cavalry mount of 1700 stood at 14 hands. By 1800, he was 14 hands 3in and by 1900, 15 hands 2½in. By and large, the smaller the horse the more stamina he had and the better he could survive on the often scant diet on campaign.

The British army dates from the Restoration in 1660, with the Life Guards and Foot Guards, literally Charles II's personal protection squad, forming the nucleus. Other regiments were added to both cavalry and infantry as events demanded. For example, the attempted coup by the Duke of Monmouth in 1685 saw an increase in mounted men to the extent that some historians claim that this is the actual beginning of the British army.

The great horses ridden in the Civil War were also called black horses and this became the colour of most British regiments' mounts. Exceptions were so rare that they entered the catalogue of regimental names – the Queen's Bays and the Scots Greys.

The late seventeenth century saw the replacement of Spain by France as *the* superpower in Europe and, since Britain and France

The Household Cavalry horse badly injured by an IRA bomb. Despite seven wounds, he survived to become a national celebrity.

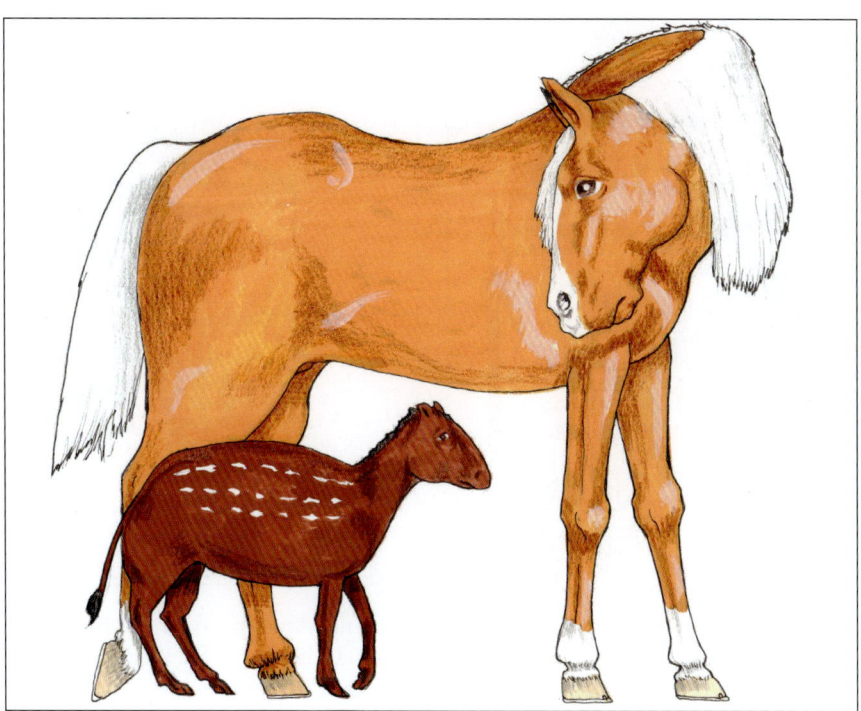

The earliest horses were about the size of dogs and were hunted by early man. This illustration compares the Dawn Horse with its modern descendant.

Alexander was the most formidable soldier of the ancient world, conquering a vast empire by the time he was 30. Bucephalus – the bull-headed – was a difficult animal that only he could ride.

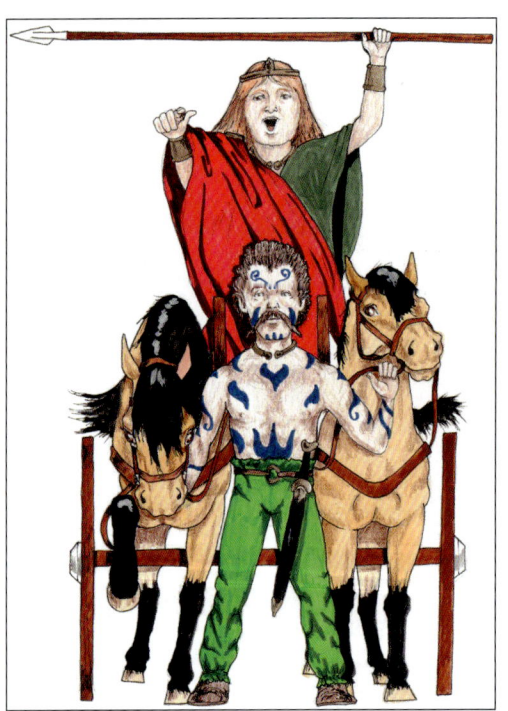

Above left: Horses seem to have been used to pull war chariots before they were ridden in battle. Julius Caesar reports that the Celts in Britain used them and they were still doing so under their 'tall and terrible' queen in 60AD.

Above right: The maddest of Rome's emperors, Caligula (meaning 'little boots') was so fond of Incitatus that he made him a consul. He was easily the most pampered horse in Rome.

Above left: The Cid was a Castilian warlord prepared to fight for his own people and the Moors in eleventh-century Spain. Babieca means booby, but the Cid saw the strength in him. He was probably an Andalusian Barb.

Above right: The Mongols were the most formidable cavalry force of the Middle Ages. They usually led one or more extra horses on the battlefield and were able to leap from saddle to saddle. The West was not ready for the advent of Genghis Khan, Subedai's overlord.

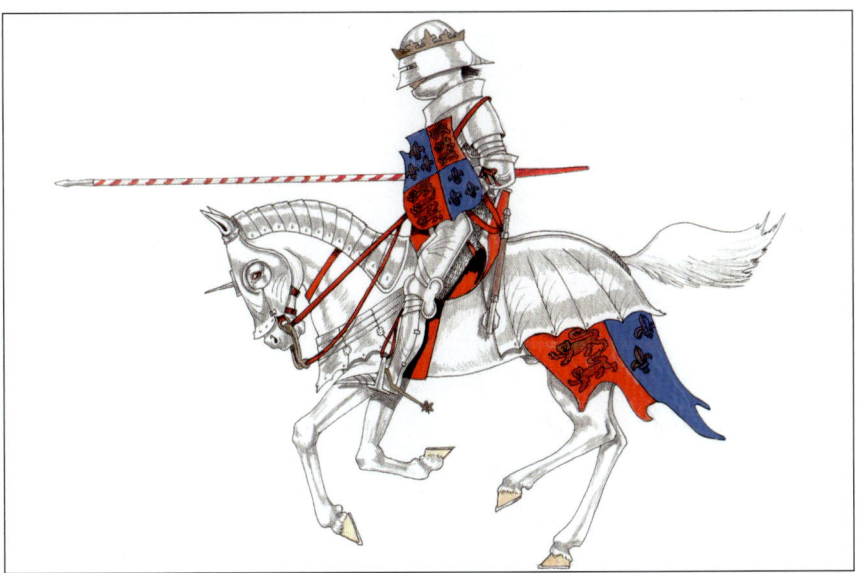

Richard was the last king of England to die in battle, at Bosworth in 1485. Surrey may have been of Arab stock, but would have had to have been a large horse to carry the large weight of armour required at the time.

Above left: The conquistadors brought two innovations from Spain – gunpowder and horses – and the Aztecs had never seen anything like them. Morzillo was bought by Cortes from one of his officers.

Above right: The most experienced of Charles I's commanders in the English Civil War, Rupert had a hell-for-leather reputation which exhausted his horsemen and ultimately failed to achieve results.

The Prussian ruler was a philosopher, poet and musician and was often seen on campaign playing his flute by the roadside. Note the short cropped tail which was achieved by removing the last bones of the spine and was fashionable in the eighteenth century.

The Emperor had a large stable of Arab horses, most of them grey. Although an Englishman bought Marengo, whose skeleton is on display at the National Army Museum in Chelsea, it is by no means certain that Marengo was actually the horse that Napoleon rode at Waterloo.

Britain's best general and one of its worst prime ministers, Wellington won immortality by defeating Napoleon at Waterloo in June 1815. At the end of a long day, Copenhagen lashed out with his hind leg and narrowly missed Wellington's head.

Not all heroic horses have heroic names! Lord Cardigan and his chestnut were the first to reach the Russian guns in the Charge of the Light Brigade in October 1854. Despite the murderous fire, Cardigan emerged with only a scratch and Ronald was completely uninjured.

Embarrassingly poor at West Point and a failed businessman before the war, Grant was nevertheless selected to command the Union army in the Civil War. Only two men ever rode Cincinnati.

Above left: Lee was the commander of the Confederate forces and one of the most able cadets at West Point. Traveller was photographed innumerable times but, in common with most commanders, he was only one of several animals that the general rode.

Above right: Most books contend that Comanche was the only survivor from Custer's command at the Little Bighorn. In fact, several cavalry horses survived because the victorious Lakota had no use for them – they could not survive on the coarse grass of the plains. Comanche was wounded and his stuffed body is still on display in the University of Kansas.

Above left: The Lakota tribe combined with the Cheyenne in 1876 to defeat Custer's Seventh Cavalry. Plains Indians did not give their horses names but, rather like their own clothing, the horses were painted with religious symbols.

Above right: The future Lord Roberts of Boer War fame was a popular commander in the Second Afghan War, 1879-80. His grey reached a kind of fame on cigarette cards in the 1930s.

Left: By the time horse and rider charged on the Western Front in the First World War, horsed cavalry was nearly obsolete; the age of the tank was about to arrive.

Below: As well as the accoutrements shown here, there would be a water canteen and a bag of hay.

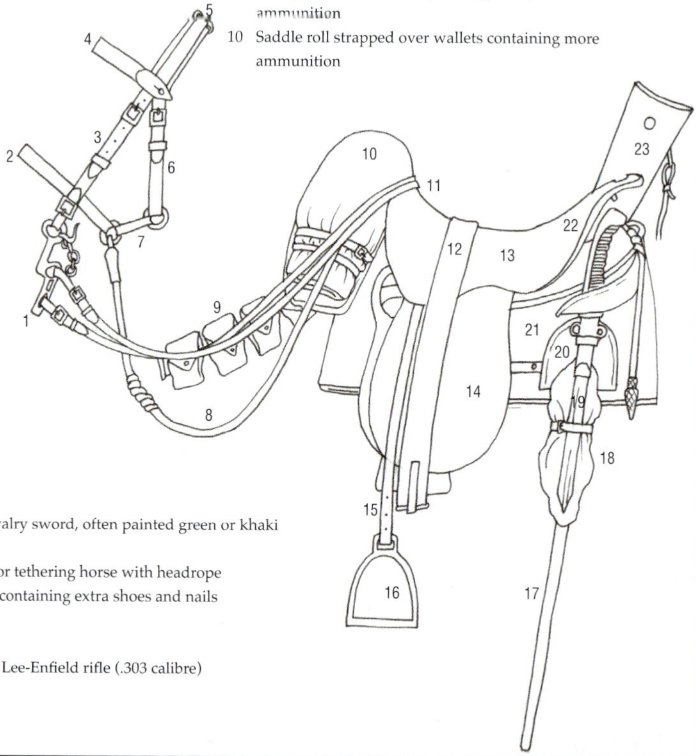

1. 1902 pattern steel bit for double rein
2. Nose band
3. Cheek piece
4. Brow band
5. Crown loop
6. Throat lash
7. Jowl piece
8. Head rope for tethering horse
9. Nine-pocket bandolier containing ninety rounds of SMLE ammunition
10. Saddle roll strapped over wallets containing more ammunition
11. Double reins
12. Surcingle/girth
13. Seat
14. Flap
15. Stirrup leather
16. Stirrup iron
17. 1908 pattern cavalry sword, often painted green or khaki
18. Watering bag
19. Picketing post for tethering horse with headrope
20. Horseshoe case containing extra shoes and nails
21. Saddlecloth
22. Cantle
23. Short Magazine Lee-Enfield rifle (.303 calibre)

had been at each other's throats since the eleventh century, a clash between them was probably inevitable. In the reign of Queen Anne, that meant the astonishing series of victories under John Churchill, the first Duke of Marlborough. The Age of Reason it may have been, but it was also the Age of the Sun King as Louis XIV established himself as the greatest sovereign in Europe.

Louis had been king since he was 5, but in 1661 when he emerged as an independent ruler free from the clutches of Cardinal Mazarin, he created the Mousquetaires Gris and the Mousquetaires Noir, 300 in all, the colours displayed on their horsecloths. When the novelist Alexandre Dumas wrote his Musketeers books in the nineteenth century, he modelled his hero, D'Artagnan, on Charles de Batz, killed at Maastricht in 1673, who commanded the Gris squad.

Although Louis spent a fortune on every aspect of government and warfare, extending his country's borders as far as he could, he understood that a cavalry regiment was two and a half times as expensive to maintain as an infantry unit. The adaptation of firearms in the late seventeenth century, with the arrival of the flintlock musket and socket or ring bayonet, meant that the footsoldier was once again master of the battlefield. The first bayonet had been plugged into the muzzle of the gun, so that it could not be fired; the socket pattern meant that a soldier could advance and shoot at the same time. The pikemen of the Civil War era had gone forever.

French cavalry, like those in Sweden and elsewhere, now settled into distinctive roles. The light cavalry, which included Croats, Cossacks and Hussars in some armies, acted as scouts riding ahead of an army, the 'eyes and ears' of a main force. They foraged and provided an armed guard for a camp. They also raided deep into enemy territory and became linked by civilians with banditry and brutality. Tsar Ivan IV of Russia (the Terrible) used his Cossacks as a private police force, terrorising enemies in his own territory as well as foes outside. These men had the heads of dogs and wolves tied to their saddles. The Horse proper were the heavy cavalry, 'big men on big horses' used for shock action, horseman against horseman or to scatter weakened and wavering infantry. There were times when the distinction between the two became blurred, especially in Britain, where Light Dragoons often charged in battle and 'Heavies' escorted kings and generals around a battlefield.

Dragoons were increasing in number and usage as mounted infantry as the seventeenth century died. By 1690, Louis XIV had forty-three dragoon regiments and sixty-six of horse. Tsar Peter the Great, anxious to westernise his Medieval state, had only one dragoon outfit in 1700 but twenty-four eleven years later. William III of England, the Dutchman who had been invited to oust James II in the Glorious Revolution, increased his dragoon regiments to eighteen from seven.

The Polish army of the period still wore armour and had large elaborate feather 'wings' attacked to their backs. Rather as the Hussars of the Hungarian army (in the pay of the Austrians) wore what had originally been shepherds' cloaks and fur caps, the 'winged lancers' equipment was impractical but was retained for its striking appearance. The lancers were confusingly called hussars, even though they carried lances and, routing the Turks before the gates of Vienna in 1683, one eyewitness wrote, 'The hussars attacked the Godless Turks like angels from heaven.' With leopard and bear skins draped over their saddles, the barbaric appearance of these men probably terrified the Turks, already a declining force in military terms. How far the riders believed that the wings had a divine protective function, rather like the war shirts of the American Plains Indians, is difficult to say. The Polish cavalry had a reputation second to none. Their watchword was 'First we defeat the enemy; then we count them.' They defeated Russians, Swedes and Turks in whirlwind campaigns throughout the seventeenth century. These troops had servants with them to carry food and spare equipment, eliminating the need for a baggage train, always the weak point for any army of horsemen.

It was unfortunate for Louis XIV that he crossed swords with one of the most brilliant generals in history – John Churchill, who became a national hero and the first Duke of Marlborough. Brought up in the tradition of Rupert of the Rhine and William Cavendish, the Duke of Newcastle who established a riding school at Bolsover Castle and created the double bridle, with curb and snaffle to control a horse better, Churchill became Louis' bête noire. He used his cavalry in close conjunction with infantry and artillery and delivered killer blows with his horsemen at Blenheim, Ramillies, Oudenarde and Malplaquet. At Elixheim in 1705, he led charges in person, even though by now, commanding officers rarely risked their lives in this way.

Charles XII of Sweden was another great cavalry commander. A relentless taskmaster, he pushed his horsemen to the limits, but always for a purpose and with astonishing success. In 1704, he pursued the defeated Saxons for nine days after which he beat them again.

In 1745, an English infantry officer, Richard Kane, wrote *Discipline for a Regiment of Foot upon Action, also the most essential discipline of the Cavalry*:

> It is sufficient for them to ride well, to have their horses well-managed and trained up to stand fire; that they take particular notice what part of the squadron they are in, their right- and left-hand men, and file-leaders, that they may know how to form. That they march and wheel with a grace and handle their swords well, which is the only weapon our

British Horse makes use of when they charge the enemy; more than this is superfluous.

While Britain and France were at loggerheads in the reigns of Queen Anne and George I, the British cavalry were undergoing reorganisation. From 1729, cavalry horses had to be 15 hands 1in tall; for dragoons, 15 hands exactly – 'strong and well-bodied'. Animals 5 or 6 years old were preferred; any younger and they were too skittish for the battlefield; any older and they would not have enough years of service left in them. The fashion by this time was for the Cadogan dock, named after Churchill's chief-of-staff and served no function at all. Horses' tails were cut by removing the last bone of the spine. Since the tail is a natural fly-wisk for the horse, this must have added to an animal's discomfort, especially in the summer. Some regiments also clipped their horses' ears on the grounds that they would sooner or later be slashed by a sword cut anyway.

In 1746, after the defeat of the last Jacobite rebellion under 'Bonnie Prince' Charles Stuart, the British cavalry were renamed, probably as an economy measure, a problem armies have faced for centuries. Three regiments were called Dragoon Guards, the oldest line regiments after the Household Cavalry. Subsequent regiments, raised in the 1790s and beyond to combat the threat from Revolutionary France and Napoleon, were dragoons and although they fought and were equipped identically, a great deal of social snobbery existed between the two. Colonels were responsible for the mounting of their regiments, officers buying their own animals. The preponderance of large black horses came from the cross-breeding of heavy hunters with the Oldenburg breed from Holland. The Oldenburg could reach 17 hands 2 inches and was descended from Andalusian/Barb cross with Medieval Friesian warhorses. Such animals were necessary because the weight they had to carry was increasing. From 1758, most European horse regiments carried valises (leather cases) tied behind the saddle for items of equipment.

Light Dragoon animals were everyday riding horses but could not be less than 14 hands 3in high. By 1760, when George III came to the throne, there were seven such regiments in the army. At Emsdorf in that year, the 15th Light Dragoons won their first battle honours in actions against German troops in the French service.

One of the most impressive exponents of cavalry in war was Maurice of Saxony, usually known as Marshal Saxe. In his *Reveries* of the 1730s (published several years later) the expert wrote that cavalry, horses and men, should be pushed hard and trained accordingly. He believed in a strict distinction between heavy and light cavalry (which the British ignored) reserving his Heavy Horse for the charge. The animals he favoured were Holsteiners, descended from the Medieval warhorses of the Gothic knights of the fifteenth century. Breed books today talk of 16–17 hands, but Saxe's horses were the tallest ever seen on any battlefield at 20 hands. His men, on the other hand, were to be slim and athletic. He despised firearms, claiming that a cavalry regiment that fired carbines was a regiment already defeated; the charge and the sword were the true business of horsemen on the battlefield. He also advocated the continued use of three-quarter armour for the riders, forty years after it had largely fallen out of use. The sword was to be straight, four feet long and used for thrusting and slashing. The front rank should carry fifteen-foot lances too and all units must stay in close formation, knee to knee, à la Cromwell and the Swedes and not to break off in useless pursuits. The pace of an attack rose from the trot at 100 yards to the gallop at twenty. He may have been wrong about armour, although such was his reputation that the French cuirassiers of September 1914 still rode into oblivion wearing breastplates.

With the rise of Brandenburg-Prussia by the middle of the eighteenth century, Frederick the Great came to dominate the battlefields of Europe. Slim, round-shouldered, a superb musician and a homosexual, he and the Empress Maria Theresa of Austria have gone down in history as enlightened despots, peculiar hybrids who espoused the philosophy of Diderot, Voltaire and Rousseau, while extending their territories at the point of a sword. Frederick's cavalry was composed of three types – the cuirassiers or heavy horse; the dragoons or mounted infantry; and the light cavalry, for scouting and foraging. Most Prussian regiments wore white tunics, a relic of the buff coats of the previous century and the cuirassiers wore breast plates, but no back plates, on the assumption that they would never turn their backs on the enemy. Frederick, who led his troops in person, took no prisoners. He expected initiative and imagination. One of his standing orders read 'An officer awaiting an attack will be cashiered'. By 1745, with an increasing awareness of camouflage in battle, Frederick officially changed the all-too-visible white for sky blue. In 1866, as Prussia led the way in German unification, the Prussians destroyed the much-vaunted Austrian army, all of them still wearing white tunics. The Prussians set up a hussar regiment in 1721 and by 1745, Frederick had eight of them, including one that formed his personal bodyguard. He also established *jagers*, cavalry and infantry units made up of woodsmen and verderers, men who knew the land and could live off it, unlike the city boys of Berlin or Hamburg. The French, impressed by Frederick, set up similar units called Chasseurs, the name for huntsmen.

The typical Prussian cavalry formation in battle was composed of cuirassiers in the front lines. Three hundred paces behind them came the dragoons, less heavily armed but still formidable. At the rear came the hussars, ready to plug gaps, worry a dithering enemy and turning an orderly retreat into a rout. Central to Frederick's cavalry training was the fast gallop over long distances of up to 2 miles. It was like watching Rupert of the Rhine in action but far more controlled and purposeful. Only in the Prussian army were the hussars used as a kind of military police, as the Cossacks were in Russia.

Perhaps the most prescient of Frederick's innovations was the creation of the Horse Artillery. The problem with cavalrymen is that they were vulnerable to sustained firepower; a volley from a wall of infantry armed with flintlocks could empty saddles faster than anything else. Three drivers controlled six horse teams and eight gunners riding their own mounts cantered alongside them. This meant that field pieces on limbers and wheels, firing cannonballs of 6lb weight, could attack *with* the cavalry rather than limbering up from behind, by which time it might be too late.

By the 1680s, long before the reign of Frederick the Great, the British cavalry had assumed the rough shape it would hold until the savage cutbacks of the 1920s. At that time, when William III became king, the cavalry were the best trained, mounted and equipped in Europe. The Queen's Regiment of Horse, the Earl of Peterborough's, the Earl of Plymouth's, the Earl of Thanet's, the Queen Dowager's, the Earl of Shrewsbury's and the Earl of Devonshire's sound alien to us today. Their colonels were men of wealth and standing with close connections to the court and they vied with each other over their regiment's appearance, horses and status, rather as wealthy gentlemen had over actors' troupes in Elizabeth's time.

The eighteenth century has been called the age of the cabinet wars, European princes playing war games with each other, in movement and counter-movement, laying sieges and abandoning them, posturing before the enemy. This is a gross over-simplification. One man who *always* avoided war if he could was Britain's first 'prime' minister, Robert Walpole. He once proudly boasted in the

Commons that there were 40,000 dead on the battlefields of Europe that year and not one Englishman.

But it was not always like that. At Dettingen in June 1743, the last battle in which a king of England – George II – fought, the cavalry lost 300 men and 600 horses.

Chapter 12

Copenhagen v. Marengo

He was known as many things. In Portugal, a grateful population called him Douro, after his crossing of their river in 1809. In Spain, he was known as the Eagle because of his hooked nose. His officers knew him as the Peer and the Beau, although he had not been a duke for long and, in his plain blue frock coat and white breeches, was probably the least flashy general in the army. The men, those who had, according to him, 'enlisted for drink' called him Arty or Nosey. All admired him. Nobody loved him.

He sat, upright as ever, in Copenhagen's saddle, where he had been for the last sixteen hours. He had risen at three in the morning in his headquarters in the village of Waterloo and had ridden out at six. In the last seventy-two hours, he had managed perhaps nine of them asleep.

Copenhagen was a chestnut, 8 years old in that memorable June of 1815. He had carried his master, Arthur Wellesley, now the Duke of Wellington, through numerous battles in the Peninsula – Vittoria, the Pyrenees, and Toulouse – the last on French soil and pointless because the Emperor had abdicated days earlier. Everybody kept a respectful distance from Copenhagen, not because he was the general's favourite mount, but because he was known to be a kicker. Wellesley, as he then was, had bought the animal from Charles Stuart, his adjutant-general. He was not fast, but his stamina was phenomenal. Wellington patted the horse's neck and murmured, 'For bottom and endurance, I never saw your fellow.'

The battle was over. The last random shots had died away. Night was falling and torches flickered over the scene of carnage. There were bodies everywhere, thousands of men from the proudest regiments who had given their all that day: George Cooke's Guards Division; the infantry squares of Clinton and Picton, the grouchy old general dead with a musket ball through his famous top hat and his head; Somerset's Household Cavalry Brigade and the Union Brigade where William Ponsonby would never lead another charge; Dornberg's Brigade; Vandaleur's; Grant's; and Vivian's. Then of course, there were the Allies – the Dutch and the Brunswickers. And, beyond the lines of light cavalry thrown out as picquets to keep the scavengers away, the thickest piles of dead of all – the French Imperial Guard under their fallen eagles. Every one of them, according to the Emperor, carried a marshal's baton in his knapsack.

Well, the French had paid for that folly now. No one had counted the dead yet and no one knew, for sure, where the Emperor was. One thing was certain; he was not the Emperor now. He was Boney again. And he would be lucky if they let him rule a farmhouse.

The aftermath of a battle is not a quiet place. The moaning of the wounded and the dying was punctuated by the occasional whinny of a horse and the echo of a shot as a soldier said a sad farewell to his mount. Some animals stood riderless, reins trailing, saddles soaked in blood. Horses that could not stand writhed on the ground, instinctively trying to get up, punctured with bullets and with missing legs. Others sat on their haunches, like dogs, whimpering in shock and confusion.

Wellington watched it all, the man who had brought nearly a quarter of a century of warfare to an end. Two hundred and fifteen battles on land – fewer at sea – and millions dead. 'I hope to God,' he thought, 'I have fought my last battle … Next to a battle lost, there is nothing so terrible as a battle won.'

For the first time in sixteen hours, Wellington rode back to Waterloo. At his headquarters, his staff officers behind him, he eased himself stiffly from the saddle. He walked behind Copenhagen, making for a bath and a bed. The animal lashed out with its deadly hind hoof, missing Nosey's head by a whisker. What an irony that would have been – the hero who had brought down the great thief of Europe, the megalomaniac who had caused all these deaths, to be killed by his companion of a mile, at the moment of his greatest victory.

In 1791, the French offered to take their revolution to any European country that might be interested. Most European countries shuddered with horror; *the Ancien Régime*, based on rank, privilege, royalty and church had stood the test of time for centuries. No one could imagine an alternative system. Yet, here it was, in Revolutionary France. *Liberté, egalité, fraternité* were fine words and if successive governments in Paris did not *quite* live up to the ideals, they were at least trying.

One by one, the powers of *the Ancien Régime* lined up against the ever more bizarre ideas of France. Nobility and church lost their powers and any politician hoping for a place in the new order had to reckon with the mob who smashed the Parisian prison, the Bastille, as a symbol of tyranny. These *sans-culottes* (who wore ragged trousers rather than the elegant breeches of a gentleman) were all too happy to loot country villas, set up tribunals with the power of life and death and build a guillotine in the Place de Grève to eliminate opposition.

Out of the chaos of the Terror in the mid-1790s emerged 'the sword of the Revolution', a Corsican artillery lieutenant called Napoleone Buonaparte. Driven by ambition and blessed with a natural military genius, he latched on to the great men of the day, first Maximilian Robespierre, then, when he fell, Paul Barras. With a handful of soldiers and acute presence of mind, he literally saved the Directory from destruction.

As a thank you, Bonaparte (he quickly dropped the Corsican spelling) was promoted from lieutenant to general; such things were only possible in Revolutionary France. He was also given command of the Army of Italy and, finding it unpaid, mutinous and ineffective, turned it into a formidable force and beat both his enemies, the Austrians and the Piedmontese, in a whirlwind series of battles which became his trademark.

The young general (he was 26) was a bad rider all his life. In Corsica, his family home in Ajaccio did not even have stables and he learned to ride with his legs dangling, slouched in the saddle. Nevertheless, he loved riding, as exercise and perhaps a chance to clear his head. He often rode 50 miles a day, much of it at speed and was often thrown. On campaign, he rode hard, pushing his men to their limits and working harder than any of them. Although by training and instinct an artilleryman, he held the French cavalry in high esteem; several cavalry commanders were elevated by him after 1804 to the rank of marshal and the most spectacular of them, Joachim Murat, became his brother-in-law and king of Naples.

Three months older than the fiery general, Arthur Wellesley was the fourth son of an impoverished Irish peer, Lord Mornington. He is attributed with saying that his greatest victory, Waterloo, was won on the playing fields of Eton, because of the high number of public school educated officers under his command that day. In fact, Wellington hated his time at the school and in his day, Eton actually had no playing fields. Using *the Ancien Régime* purchase system, Wellesley bought a commission in the 73rd Foot at the age of 16 and indiscriminately exchanged through five regiments before taking command of the 33rd (later called the Duke of Wellington's Regiment) in 1793, the year in which William Pitt's British government declared war on Revolutionary France. Like Napoleon, Wellesley was a political officer. He became a member of the Irish Parliament in Dublin while still a lieutenant. Unlike Napoleon, Wellesley had little time for the British cavalry, complaining in the Peninsula that they got him into scrapes.

In 1798, Napoleon had sufficient 'clout' to choose his next campaign. Anxious to create an empire for himself and to break Britain's link with India, he led an expedition to Egypt in July. He defeated the Turkish rulers twice in the following year and the Mamelukes with their wild cavalry mounted on magnificent horses. The riders lacked discipline, but their elan was impressive and Napoleon introduced them into the French cavalry. They escorted him on battlefields without number and one of them, Roustam, slept outside his private rooms at Malmaison or his tent on campaign, armed to the teeth. Such was the panache of these horsemen that their ivory-hilted curved sabres were the 'must have' weapon of French and British light cavalry. Napoleon and Wellington both carried one and they continue to be the dress weapon of generals in the British army to the present day. Rather bizarrely, US Marine officers wear them too.

The downside of Egypt was that the French fleet was destroyed by some dazzling (and risky) tactics from Horatio Nelson at Aboukir Bay; Napoleon had to abandon Egypt and go home by land.

In the coup of Brumaire (the Revolutionary government had replaced the old calendar with a series of new months – this one was 'foggy' for November) Napoleon overthrew the Directory and established the Consulate, with him as First Consul for life. What followed was two years of warfare during which he expanded his army, built up an impressive stable of grey Arab horses for his own use and achieved the peace he needed to consolidate his political position.

Wellesley's career took him away from Europe during those years. His brother, Lord Mornington, was Governor-General of Bengal, one of the presidencies of the East India Company (see Chapter 16) and he pulled strings to bring Wellesley's regiment out to India. He defeated the sultan Tipu Sahib at Seringapatam in the year that Napoleon seized power and broke the ascendancy of the Mahratta princes at Assaye and Argaum four years later. By 1805 he was back in England, known, somewhat scornfully, as a 'Sepoy General'.

There had been a brief interlude in 1802 with the Peace of Amiens when the British aristocracy and gentry flocked to Paris to see what the First Consul had made of his capital. They were disconcerted to find that several of them were caught by the outbreak of war and unable to leave France for the next ten years – the longest 'grand tour' in history!

British policy had always been to fund coalitions of the Europeans against Napoleon and this continued, but time and again, the French would destroy individual armies piecemeal and the coalitions collapsed. By 1807, Napoleon had crowned himself Emperor of the French, in the vague hope that that title would give the impression of his being elected – an emperor crowned with a ballot paper. The truth was that the Emperor was another in a long line of French autocrats, issuing bulletins (which were known to be lies) on a regular basis. But as long as he kept winning victories, which he did, the French basked in 'la gloire' and were happy under his rule.

The Archduke Charles of Austria whinged that French cavalry were poorly mounted, bad riders and badly equipped. The reason that they won battle after battle, he claimed, is that they charged everything with a fighting spirit lacking in other armies. In battle, the cause is important – the ideals of Revolution or adoration of the Emperor – but it cannot make consistent victors out of bad horsemen.

Other than a stubborn wish to overthrow 'Boney' and kick French backsides (something they had been doing for eight centuries) the British cavalry had little to spur them on other than loyalty to their regiments. The heyday of the French cavalry was 1805–07, with brilliant victories at Ulm, Austerlitz, Jena, Eylau and Friedland. The cavalry made up 10 per cent of the army and was well-mounted, with experienced troops not reliant on foreign support. That said, both the Mamelukes and the Polish Lancers of the Guard were valuable imports to the army.

There were seven categories of cavalry: the Heavies were made up of Cuirassiers, Carabiniers, Dragoons, Grenadiers à Cheval and Gendarmes d'Elite. The Cuirassiers were formidable. Their horses were 17 hands high, bred in Normandy and their riders were selected for their size and experience. Napoleon favoured them over all other categories and this was reflected in their extra pay – five centimes more than other units, the 'sou of the grenade'. Known as Napoleon's *gros frères* (big brothers) they were kept in reserve on the battlefield, ready to smash through exhausted infantry. With their leopard-skin turbaned copper helmets and back and breast plates, the Cuirassiers looked impressive, but there where those who complained that their armour was too heavy and that it slowed their charges to a canter at best. They carried long, straight swords.

The Carabiniers were named after the short muskets they carried on the battlefield. Their horses were always black and they had a fierce rivalry with the Cuirassiers. Their association with royalty went way back, although no one could doubt their loyalty to the Emperor.

The Dragoons, by 1800, had shaken off the reputation of mounted infantry and fought in formation as cavalry. Their helmets were in the Grecian style with a long horsehair plume which could cause infection if caught up in wounds and they wore a tailed tunic (*habit*) of dark green.

The Grenadiers à Cheval had the swagger of Cuirassiers and pay almost as good. Other units called them, with a hint of jealousy, 'the gods' and 'big heels' and 'giants'. Their straight swords had a grenade design in the hilt. Captain Mercer of the British Horse Artillery was impressed by them at Waterloo, although their numbers had declined dramatically by that time.

The Gendarmes d'Elite were the military police, made up largely of the gendarmerie of Paris. They had been raised in 1802 after various attempts to assassinate Napoleon and, along with the Mamelukes, they formed the emperor's bodyguard after 1804. They did not come under fire often and earned the nickname the 'immortals'. Their reputation was formidable, especially among would-be deserters and pillagers caught with loot in their possession.

The French light cavalry was made up of the Chasseurs à Cheval, Hussars, Lancers and Mamelukes. The Chasseurs were formed in 1779 and in keeping with all light cavalry, they acted as scouts, foragers and raiders. Napoleon expected perfection from these men. In 1812 he wrote to Louis Berthier, his chief-of-staff, 'A colonel of chasseurs or hussars who goes to sleep, instead of spending the night in bivouac and remaining in constant communication with his picquets, deserves to be shot.' The Chasseurs carried curved sabres and short carbines, often acting as skirmishers ahead of an assault. The Guard Chasseurs were known as 'the invincibles' and 'the cherished children'.

The Hussars were the most flamboyant of the French cavalry, contributing in no small measure to the huge cost of Napoleon's campaigns. With their mirliton caps and flounces, their braided dolmans and pelisses and their tight breeches and Hungarian boots, they had a reputation as *beau sabreurs* and womanisers. They were as careful with their braided hair as the Spartans had been in the ancient world and their reputation went before them. The Hussar general Antoine Lasalle famously wrote, 'A Hussar who isn't dead at thirty is a blackguard.' He himself was killed leading a charge at Wagram in 1809; he was thirty-four. The curved sword the Hussars carried was known affectionately as the 'bandy one', but most officers preferred the Mameluke pattern. So impressive were Lasalle's horsemen in the Jena campaign against Prussia that Napoleon said to the general, 'If your light horsemen can capture fortified places [Stettin] in this way, I shall have to disband my engineers and melt down my siege guns.'

Lancers had an exotic appearance inherited from their Polish originals. The Polish Lancers under Josef Poniatowski, one of Napoleon's marshals, wore the distinctive czapska headdress with its square top. The French regiments wore brass helmets of dragoon pattern and it was only the lance, sometimes carried by the front rank only, that told them apart from dragoons. The lance was ten feet long and weighed 7lb, with a wooden staff and steel point from which fluttered a scarlet and white swallow-tail pennon. The weapon had its detractors, but, reaching four feet beyond a horse's head, it outranged musket/bayonet and sword and, with the weight of a man and horse behind it, was a formidable deterrent.

The Mamelukes, created in 1801, were composed of natives from Egypt, Syria and later, at least twelve other countries. They fought superbly at Austerlitz and were given their own eagle standard as a reward. Their razor-sharp swords could decapitate an enemy and they sharpened their stirrups, using the edge to keep horses away and hit an unsuspecting infantryman in the face.

Twenty-four years of fighting and 215 battles covers thousands of miles and necessitates a whole industry geared to the feeding, grooming, stabling and training of horses. Off the field, cavalrymen wore short shell jackets, forage caps, canvas trousers and wooden clogs. Horses were exercised twice a day, unsaddled and rubbed dry. A curry comb removed dirt from the animal's coat because a horse breathes through its pores as well as its nostrils. A dandy brush and wisps of straw were used for the final 'polish', especially for regimental parades and field days. Each horse was fed 10lb of hay a day, 15lb of straw and ⅔ of a bushel of oats crushed to a mash. A little rock salt or treacle was sometimes added as a treat. The harness then had to be cleaned and hung up for inspection by the colonel or troop captain. The light cavalrymen had Hungarian-style bridles and double straps with a cross-piece over the forehead. A metal crescent hung below the animal's throat which could be used to tether the horse on campaign. The saddle, again Hungarian or high-mounting, had a wooden tree and pigskin seat and pilches. It fitted over a blanket to prevent sores, especially when an animal might be ridden 30 or 40 miles a day. Over the saddle, a cloth shabraque was worn, with the regimental cypher on the hindquarters. These were sometimes left in depots and barracks and sheepskin took their place. The hussar pattern stirrups were rounded and most units carried a valise behind the saddle which contained cloak, spare boots and other items. In front of the rider, lashed to the saddle's pommel, was a holster containing a pistol (rarely used) and a pouch for horseshoes and nails. Every man was expected to be able to shoe his own horse.

Training cavalry horses under Napoleon was hard work. A horse will not instinctively run into danger or anything that frightens him and this has to be overcome as far as possible to make cavalry effective. While horses were taught the tricks of the parade ground, like the Lipizzaners of the Vienna Riding School – the piaffe, the capriole and the levade, much of it was getting used to the explosion of cannon and musket, the shouts of fired-up men and the waving of flags. Napoleon's groom, Jardin, drove pigs and dogs between the legs of his master's horses to get them used to the unexpected.

Napoleon's own horses are complicated. The skeleton on display at the National Army Museum is that of Marengo, one of several grey Arabs that the emperor rode. He passed through a series of owners after his capture at Waterloo and achieved a kind of immortality as the charger of the man whom von Clausewitz called 'the god of war'. Unfortunately, Marengo is consistently absent from the meticulous record that Napoleon kept of his horses. There are umpteen portraits of Napoleon, from his early days on the Italian campaign to his final defeat. In most of these he is mounted on a grey and the animal is usually assumed to be Marengo. The most commonly listed horse is actually Ali, but it may be that this animal was renamed Marengo at some point or that Ali was simply the emperor's pet name for him.

We do have a list of Napoleon's horses on Elba, where he was sent in exile after his abdication in 1814. Wagram was the grey he rode at that victory; his original name was Ingenue. The emperor often referred to him as 'mon cousin'. Montevideo was a bay from South America who carried Napoleon in his brief Spanish campaign in 1808. Emir was a chestnut with dark mane and tail that the emperor rode into Madrid; he was one of several who took Napoleon on the disastrous Russian campaign of 1812 and saw service at the bloodiest battle of all Napoleon's adventures, Borodino. Gonsalvo was a large bay whose bridle was cut in two by a musket ball at Brienne in the hectic last days before the abdication. Roitelet had a series of narrow misses too; at Lutzen, the Anglo-French cross-breed chestnut had skin torn from his hock by a cannonball. He was a skittish animal, on record as having thrown the Emperor twice while on review. Tauris was a dapple-grey given to Napoleon by Tsar Alexander I while the two were still on speaking terms. He was probably the emperor's principal charger on the Moscow campaign and features in Ernest Meissonier's famous painting of the French retreat through the endless Russian snow, many years later. He is No. 902 in the official horse list in December 1809 and may be the 'Marengo' captured at Waterloo. Intendant was a pure white Norman stallion kept mostly for ceremonies. The troops knew him as Coco and the cry 'Voila Coco!' was as common on campaign as 'Vive l'Empereur!'

Accounts of individuals and their horses are rare, but we know that Private Melet of the Empress's Dragoons of the Imperial Guard had a mount named Cadet. He had ridden the animal in Prussia, Poland, Spain, Austria, Russia, Saxony and France and fought twelve major battles and dozens of skirmishes in his saddle. On the Moscow campaign, Melet had crept behind Russian lines to steal forage for Cadet and brought back a prisoner at the same time. Fittingly, Waterloo in June 1815 was the end for them both; Melet was wounded and Cadet killed.

The British cavalry had inherited Prince Rupert's reputation for riding too fast and too far, not controlling their charges and not keeping a reserve. They were officered by men who rode to hounds on tall thoroughbreds and chased the French in much the same way as they chased the fox. In India and even in the Peninsula, Wellesley had relatively few cavalrymen and fought with the instincts of an infantry commander.

The heavy cavalry was composed of the Life Guards and the Royal Horse Guards, all on black horses, but without the armour of their French counterparts. Below them, in terms of status, came the Dragoon Guards and Dragoons. They wore scarlet jackets (except for the Royal Horse Guards who wore dark blue) and 'Roman' helmets with black wool crests and plumes. They carried long, straight swords with murderous hatchet points and carbines strapped to their saddles. The regiment that stood out among the Heavies was the 2nd North British Dragoons, known as 'the Scots Greys' on account of their grey horses. Their headgear was the fur cap of the grenadier regiments which, while flamboyant, gave far less protection than the conventional helmet.

The light cavalry was composed of light dragoon regiments and, from 1805, hussars. There were no lancers in the British army until 1816. Ironically, the most impressive light unit was probably the King's German Legion, black-uniformed cavalry from Hanover, stationed in Britain and serving with the British army. While it was standard practice to pay lip service to the notion of 'horses first and men second', the KGL actually carried that out, seeing to the watering, feeding and grooming of their animals after a day's march before looking after themselves. The carbines carried by the Lights were the Eliott pattern, invented in 1773 and largely obsolete as a cumbersome piece of equipment by the time of the Peninsular War (1808–14) and the considerably better Paget version, which was shorter, more serviceable and was carried on a soldier's shoulder-belt from a swivel-hook. The curved sword of the Lights was the 1796 pattern designed by John Gaspard Le Marchant who was killed at Salamanca in 1812. This weapon caused official complaints from French army doctors because of the axe-like impact of the blade.

The training given to men was complicated, but it was never carried out by units larger than a regiment and was often confined to a squadron (two troops). Above all, it was performed on a flat, level parade ground and bore no relation to actual warfare at all. Not until 1853, just before the Crimean War, did a handful of British cavalry regiments take part in manoeuvres roughly approximating to battlefield conditions. Closed column, single line, double line, echeloned line, checkered line, all had to be mastered. So did the cuts of Le Marchant's handbook, including a bizarre parry to protect the back which few men ever mastered.

As Napoleon's political career took over, his generalship declined. He left battles and sometimes whole campaigns to subordinates and none of them could match his military genius. He went into Spain in November 1807, driving back General John Moore's British army that had been sent to support the Portuguese. He then went home, leaving it to Soult, Davout and Massena, marshals all, to wrap up. Unfortunately, they came up against Arthur Wellesley.

The British were outnumbered in the Peninsula, but command of the sea (most of the French fleet had been destroyed at Trafalgar in October 1805) meant that Wellesley could supply his army easily whereas the French had to drag their supplies across the Pyrenees. If threatened by a larger force, he retreated to the impregnable lines of Torres Vedras and waited for the right moment to venture out. A long list of victories, including the cavalry action of Sahagun, drove the French out of Spain and across the mountains where he beat them again at Toulouse on 10 April 1814, days before Napoleon left for Elba.

If they thought it was all over, it was not. Napoleon had fought a whirlwind campaign – eleven battles in four months – the last fought outside Paris. In exile, bored, bitter and in denial, he maintained correspondence with key individuals and was allowed to retain a personal bodyguard of loyalists. He landed at Frejus and most of France welcomed him back. The exiled monarch, Louis XVIII, hopelessly ineffectual, became exiled again and Napoleon marched into Belgium, preparing to use his old tactic of dividing his enemies and beating them individually. No less than seven armies had defeated him at Leipzig in 1814, but only two of them – the British and the Prussians – were close enough to stop him now.

He stopped Gebhard Blücher's Prussians at Ligny on 16 June and Wellington's British at Quatre Bras on the same day. Had Blücher done what the Emperor expected and gone home (he was after all in his seventies) there is little doubt that the battle of Waterloo, fought on 18 June, would have had a very different outcome.

Thousands of books have been written about the battle, ending, as it did, a quarter century of almost unremitting warfare. For our purposes, two actions stand out, exposing the vulnerability of both the British and the French cavalry. The first was the charge of the Union Brigade (so called because of its Scots and Irish regiments) which is usually depicted as an insane, hell-for-leather charge. Much of the blame for the misunderstanding of this manoeuvre must lie with Lady Butler, the wife of a general, whose paintings of military actions were widely admired in the 1870s. 'Scotland Forever' shows the Scots Greys thundering at the viewer out of the canvas at full gallop. What it does not show is that, moments earlier, Highlanders in packs and kilts were accompanying their mounted comrades, hanging on to the stirrup leathers. Dangerous though this was, it was possible given the fact that the Brigade rode uphill against D'Erlon's I Corps and could only have negotiated the wet and difficult ground at a trot. Nine squadrons of heavy cavalry, more than 1,300 sabres, must have been horrific to the French infantry facing them. D'Erlon's Corps, perhaps 11,000 men, reeled and were driven back, Sergeant Charles Ewart of the Greys capturing the eagle standard of the 45th Line and receiving seventeen cuts for his pains. Commissioned on the battlefield, Ewart retired later on a pension of 5s 10d a day and died a hero in 1846.

It was what happened next that caused the problem. Major de Lacy Evans, who would later be a general in the Crimea, remembered, '... our men were out of hand. The General of Brigade [William Ponsonby] ... and every officer within hearing, exerted themselves to the utmost to reform the men, but ... our efforts were abortive.' Corporal Dickson reached the cannon of the French Grand Battery with the Greys – 'We sabred the gunners, lamed the horses and cut the traces and harness. I can hear the Frenchmen crying "Diable!" when I struck at them and the long drawn out hiss through their teeth as my sword went home.'

Two and a half thousand French cavalry, heavy and light, crashed into the scattered ranks of the British, horses blown, men dazed with the exhilaration of the charge. The 3rd and 4th Lancers slaughtered as many of the Union Brigade as they could reach. Ponsonby was skewered by a lancer called Urban who dismounted and helped himself to the general's sword – a spoil of war that was returned to the Ponsonby family many years later. The Union Brigade had lost 612 men killed and wounded, 46 per cent of its strength.

At about four o'clock the second cavalry debacle took place. Wellington had lost the use of two cavalry brigades, but it was nothing to the loss that would befall the French. Michel Ney, the 'bravest of the brave', misread Wellington's order to pull his units back out of artillery range. To Ney, through thick smoke and bad light, it looked like a general retreat – the exact moment to unleash the Cuirassiers on a wavering enemy. Nearly 9,000 horsemen were drawn up, eight regiments of cuirassiers who had seen little action so far and ten squadrons of light cavalry under Lefebvre-Desnouettes. They advanced at a walk, terrifying the British infantry facing them. 'Not a man present,' wrote Ensign Rees Gronow of the 1st Foot Guards, 'could have forgotten in after life the awful grandeur of that charge. [It was] an overwhelming, long moving line, which, ever advancing, glittered like a strong wave of the sea when it catches the sunlight ... the very earth seemed to vibrate beneath their thundering tramp ...'

The British formed square, the traditional method of defence against cavalry. In the centre were the officers and the colours and the square bristled with bayonets and the muzzles of muskets, men kneeling and crouching, firing and reloading in serried ranks. Later artillery technology would render this tactic obsolete – an entire regiment could be destroyed by high-explosive shells – but the bouncing cannonballs of Napoleon's day constituted no such threat. Ney had five horses killed under him that day, three in this attack. He came through it unscathed, but three-quarters of the French generals leading the charge were killed or wounded. Not a single square broke in the British lines and the cavalry soon became hopelessly confused, riding round and round them. The odd lancer may have been able to do some damage but swordsmen could not get close enough. The order in the squares was to aim low, to kill the horses, causing the maximum chaos in the charging lines. The ground was soft after the night's rain even in late afternoon. Dead horses littering the field slowed the attack and wounded ones, panicking and in pain, bolted in any direction other than forward, taking their riders with them. Gronow wrote later, 'The horses of the first rank of cuirassiers ... came to a standstill, shaking and covered with foam, at about twenty yards distance ...'

The French cavalry were well supported by their artillery and it was the guns, not the horsemen, that did most damage to the British squares. Conversely, even though Ney's troops rode through at least sixty cannon, not one was spiked to make them unfirable. All cavalry officers had prickers on their pouch belts, pieces of silver that could be rammed into cannon touch-holes and broken off, making the gun useless. Above all, Ney did not follow up his attack with infantry. All this was based on the assumption that the British were wavering and that the mere sight of so many horsemen charging them would turn a retreat into a rout. Various historians have claimed that had Joachim Murat been there instead of Ney (the man had abandoned Napoleon by this time) the outcome would have been different. Certainly, Murat was a force of nature, delivering devastating charges at Jena, Friedland and Eylau; but if anything, he was more reckless than Ney and nothing would have been different.

The battle of Waterloo was not lost by the impetuous charge of the French cavalry. Had Marshal Emmanuel Grouchy kept Blücher's Prussians away from the field, it might well have been another victory for Napoleon. Whether he could have stayed in power with so many nations hostile to him is debatable, but the last word has to be with Wellington – 'It has been a damned nice thing – the closest run thing you ever saw in your life.'

The horse known as Marengo, whatever his name actually was, was shipped to England, although exactly when and where he was kept, is unknown. The Emperor's gilded coach and its horse team were on show at the Egyptian Hall in London's Piccadilly; 10,000 people a month flocked to see it at a shilling a pop. The poet and rebel Lord Byron had a replica made for himself, such a fan of Napoleon was he.

Perhaps for the first time in British history, there was a semi-official recognition of the heroism of warhorses. Astley Cooper, the most famous surgeon of his day, was a board member of the Veterinary College. He bought twelve badly wounded cavalry horses and removed the grapeshot personally from their bodies. As they recovered on his country estate, they could be seen in the fields forming line, advancing in line and echelon, as though they heard the bugles calling.

In 1823, two years after Napoleon had died, probably of pancreatic cancer, on St Helena, Marengo turned up at an exhibition at the Waterloo Rooms in Pall Mall. Where he had been for the past eight years is anybody's guess and there was no actual proof that this was the emperor's horse at all. The advertising poster read:

> Bonaparte's white Barb charger, Marengo, has been inspected by many of the Nobility and Gentry ... This beautiful ... Charger ... was the favourite Horse of the late Emperor and accompanied him through most of his Battles. He has five Wounds which are visible; and a bullet remains in his Tail. The Imperial crown and the Letter N are branded on his hind Quarters. He is so gentle, that the most timid Lady may approach him without fear. The superb Saddle and Bridle and the Boots that Napoleon wore at Moscow are likewise shown. The Person who exhibits the Horse is well acquainted with the Movements of the late Emperor and speaks six different languages.

No chance that he was a con-man, then!

By the mid-1820s, Marengo had been bought by the Angerstein family, financiers in the City, and was put to stud at New Barns, near Ely in Cambridgeshire. The animal died in 1832 and was articulated to be put on public display. He is still there, minus two hoofs, in the National Army Museum in Chelsea. The skeleton has recently been restored, cleaned and moved to a new exhibition in the Battle Gallery. With head thrust forward and one front leg raised, the idea is to recreate the excitement and grace of a cavalry horse in action. Since it is a collection of bone, however, it all looks rather silly.

Copenhagen lived a peaceful existence at Stratfield Saye, Wellington's country estate in Hampshire. While his master went on to become a formidable *éminence grise* of the British army and one of the worst prime ministers in history, Copenhagen wandered the fields, being spoiled and fed his favourite bread every day by the Duchess. He died in 1836 at the age of 28 and was buried with full military honours on the estate. Wellington refused point blank to have him stuffed or articulated. The horse's headstone reads:

> Here lies Copenhagen, the charger ridden by the Duke of Wellington the entire day at the Battle of Waterloo.
> Born 1808, died 1836.
> God's humbler instrument, though meaner clay,
> Should share the glory of that glorious day.

Chapter 13

Ronald

He had been in the saddle since dawn and the day was still cold. To that end, he had worn his pelisse buttoned. Although it did not occur to him often, James Brudenell, the 7th Earl of Cardigan, was 56 years old and in this God-forsaken country, the ground mist and the night frosts played hell with his weak chest.

Cardigan half turned to check his brigade's lines. At his shoulder sat the brigade-major, George Mayow. All right, the man was a 'Heavy' and he was not well, but he knew his business. Cardigan's ADC was waiting to his left; Henry Maxse, an infantryman, it was true, but efficient enough. The brigadier could not help but smile at Billy Brittain, his Orderly Trumpeter. The lad could blast out reveille and all the rest with the best of them, but how would he fare with black powder spattering his face and grapeshot singeing the air?

And they were all about to find out, Cardigan was sure of that. Those damned Heavies had charged already, General Scarlett's Three Hundred slicing right through the mass of grey-coated Russians. Then the galloper had come barrelling down from Lord Raglan and his headquarters staff on the Sapouné Heights – Louis Nolan, the cavalry 'expert' with his published books and his contempt for the high command. He had given Raglan's written order to Cardigan's waste-of-space brother-in-law, Lucan, and the men could be heard arguing all across the valley.

'What guns?' Lucan had barked. 'Where away?'

Nolan had flung out an arm without even looking behind him. 'There, my lord. There is your enemy. There are your guns.' Lucan had looked. Cardigan had looked. All they could see, at the end of the North Valley, perhaps a mile and a half away, was a Russian battery in position, a cul-de-sac of death. Surely, Raglan could not mean Cardigan's Light Brigade to charge *them*? It was suicide.

Cardigan had pointed out as much to Lucan. Detest each other though they did, they were cordial this morning. 'Certainly, sir,' Cardigan had said, 'but allow me to point out to you that the Russians have a battery in the valley at our front and batteries and riflemen on each flank.'

'I know it,' Lucan had fumed, still seething at Nolan's insubordination, 'but Lord Raglan will have it. We have no choice but to obey.'

Then, the incompetent buffoon who commanded the cavalry showed his true colours. Lucan interfered with Cardigan's command, pulling the man's old regiment, the 11th Hussars, back to form a second line and the 17th Lancers, Lucan's old regiment, to the front. Cardigan had swung his chestnut, Ronald, around and trotted back to Lord George Paget, sitting his horse at the head of his 4th Light Dragoons – 'You will take command of the second line and I expect your best support – mind, your best support.'

Back in position at the head of the brigade, Cardigan drew his Mameluke-hilted sword. The scream of steel brought Ronald's head up and he shook his mane as his rider angled his heels down and gripped with his thighs.

'The Brigade will advance,' Cardigan barked, the familiar voice of 'Jim the Bear' clear in the otherwise silent valley. 'First squadron of the 17th Lancers direct.' Cardigan glanced across. At the head of that unit sat Captain William Morris, who had already fought three battles like this one in India. But Cardigan had nothing but contempt for 'Indian officers' and he had already had words with Morris this morning when the young puppy had the barefaced effrontery to urge Cardigan to charge to support the Heavy Brigade. And there was that other insolent blackguard, Nolan, sitting beside him. Had the world gone mad?

'Well,' Cardigan muttered to himself, 'Here goes the last of the Brudenells,' and he nudged Ronald forward with his spurs.

Brittain's trumpet behind him blasted out the advance and the line moved forward, the pounding of hoofs a backdrop to the jingling of bits and clanking of scabbards. The lances of the 17th were pointing skyward, their scarlet and white pennons snapping in the breeze. Cardigan and Ronald were three horses' lengths ahead of his brigade, exactly the right position according to the cavalry manuals and the brigadier had no idea what a target he made, glittering in his braided hussar jacket and crimson overalls, riding a fine horse with white stockings.

Suddenly, from the left, Cardigan saw a rider cutting across his front. It was Nolan, the staff officer. What the hell was he doing? Cardigan could barely make it out but it seemed as if Morris, as alarmed by this as everybody else, shouted out, 'That won't do, Nolan. We've a long way to go and must be steady.'

A shell screamed across the valley, the first of many, and Nolan's troop horse wheeled round, his rider's sword gone from his hand. The animal, confused and panicking, cantered back through the lines of the 17th and Cardigan vowed to have the man court-martialled. *Nobody* rode ahead of the leader of a brigade – the very idea!

Cardigan swept on, relaxing the reins as Ronald moved from the walk to the march to the canter and finally the gallop. The wind roared in the ears of horse and rider who both saw Russian artillery at the end of the valley hauling their guns forward and firing them. Then it was all smoke and a trembling ground, Ronald's ears flat, his eyes wide and his teeth bared. Cardigan stayed upright in the saddle, steady as a church, his sword blade at the slope on his shoulder.

He was leading his men, according to Alfred, Lord Tennyson, who read William Russell's eyewitness account in *The Times* weeks later 'into the Jaws of Death, into the Mouth of Hell'.

Of the 678 men who rode the charge of the Light Brigade, 103 were killed and 127 were wounded. Among the horses, the figures were even grimmer; a conservative estimate gives 362 killed or shot later for wounds.

Not all heroic horses have heroic names and perhaps it is fitting that one of the worst-handled cavalry actions in history should have been led by an over-promoted idiot riding a horse called Ronald.

Astonishing victory though it was by the British and Prussian armies at Waterloo, what followed, at least from the British point of view, was the 'long peace', military inactivity in which the army rested on its laurels as the force that had beaten Napoleon and failed to keep up with either technology or the march of progress.

As we have seen, the officer class was made up of gentlemen, particularly in the cavalry where snobbery was a way of life. In the cavalry and the infantry, a young man bought his commission, with the rank of cornet and settled into a life of parade-ground drills, balls and hunting which had very little to do with the 'bloody trade' of soldiery. There were regulation prices laid down by the War Department (the Horse Guards along Whitehall) in a system of purchase which had first been laid down in Queen Anne's reign. A cornet could expect to shell out £840 for his commission in the 1840s; a lieutenant would have to pay £1,190; a captain £3,225; a major £4,575 and a lieutenant colonel £6,175. Certain fashionable regiments, like the Life Guards and the Royal Horse Guards, cost considerably more and such was the competition to find a post in an elite unit, that over regulation prices were constantly being paid. James Brudenell, the 7th Earl of Cardigan, who rode Ronald in the charge of the Light Brigade, paid over £35,000 for the lieutenant-colonelcy of the 15th Hussars and more than £40,000 for the 11th. Since a commission was a man's personal property, the whole career of a cavalry officer was actually a business transaction. It meant that wealthy men with no aptitude at all could ride roughshod over talented junior officers who were too poor to climb in the ranks. This sorry state of affairs was not rectified until 1871 and even after the abolition of the purchase system, there was a tendency for titled young men to join the cavalry; nearly 50 per cent of the 17th Lancers' officers in the First World War came from aristocratic families.

On top of the crippling expense of a commission and promotion, a cavalry officer had to equip himself *just so*, often at the personal whim of his commanding officer. The light cavalry in particular, many of which regiments were equipped as hussars after 1815, wore exotic uniforms with yards of gold lace, dozens of buttons (a hussar's jacket had ninety-six) and all of it was provided by London-based tailors who did not come cheap.

Horses too were turned out in dazzling finery. The double bridles had cross-pieces over the forehead, bits decorated with regimental bosses, a throat-plume of coloured horsehair. Over the, usually privately purchased, hunting saddle, was a shabraque edged with gold lace and adorned with the queen's cypher and the battle honours of the regiment. On top of that was a sheepskin or leopard-skin pilch, all held in place by a surcingle. Even though much of this finery was left behind on campaign, it all had to be bought, maintained and stored, which carried ongoing costs which soon mounted. Lieutenant Winston Churchill, joining the 4th Hussars in the 1890s, took six years to pay his tailors' bills.

By contrast, the Other Ranks in the British cavalry often enlisted through hardship rather than for career prospects or an exciting adventure. The majority of men who joined the cavalry in the 1840s and '50s were labourers, brought up on the land, with limited special skills. Large numbers of them were Irish, driven off the land by the failure of successive potato harvests from 1845. Better-paid work could be found on the new railways (to which the army, specifically the Duke of Wellington, had an aversion) but it was hard and dangerous. Many men opted for the two meals a day, cheap beer (pongolo) in the mess canteen and a free suit of clothes, replaced every Christmas.

The average age for a cavalry recruit was twenty-one. Louis Nolan, the expert who was the first man killed in the charge of the Light Brigade, wrote that a light cavalryman's ideal height was 5ft 4in, but most of them were considerably taller. Nolan's friend William Morris, known as the 'Pocket Hercules', was strong and powerful but stood at 5ft 7in, the actual average height for the cavalry. A private enlisted for twenty-one years and could be expected to serve both overseas, perhaps in India or Ireland; and to 'aid the civil power', that is, support the police, as the 15th Hussars did at 'Peterloo' in Manchester in the summer of 1819.

Most ordinary soldiers could not ride, unlike their officers who rode to hounds from an early age. Even in cases where recruits could ride, they were taught the 'army seat' by the regimental riding master and his team of rough-riders, whose job it was to break and train the horses too. The best animals were thoroughbreds from Ireland, but most counties had horse fairs and the army bought there, along with civilian farmers. Louis Nolan could train a troop horse in sixty-four days, but the norm was ten months, beating drums and firing guns near them to get them used to battle conditions. Bearing in mind that regiments did not play 'war games' until 1853 (and then, not many of them) it is arguable that the horses were better trained than their riders. Only in the Curragh and Phoenix Park, Dublin, was there space for open cavalry formations as opposed to parade-ground drill.

These formations were complex. There was column, direct echelon, oblique echelon, pivot, attack and charge, all directed by bugle calls. The pace of movement was the walk, at 4 miles an hour, through the march, trot, canter and gallop at 11 miles an hour. Soldiers were taught to ride with legs straight, rather as Medieval knights were. This enabled them to use sabre or lance effectively but served less purpose when firearms were included. On paper, each soldier carried a carbine, a short musket (later rifle) designed to be fired with one hand from the saddle. In reality, these guns were difficult to load and were not often used, particularly in battle. Stubborn tradition played its part here. When Edward Hodge, colonel of the 4th Dragoon Guards, was offered revolvers for his men in the Crimea, he refused, believing them to be dangerous and new-fangled.

In the lists of officers compiled by the War Office, Veterinary officers rank higher than other 'specialists', even surgeons. This is perhaps not too surprising in a country that set up the Royal Society for the Prevention of Cruelty to Animals in 1824 but waited unto 1881 for the National Society (no royal patronage here) for the Prevention of Cruelty to Children.

Life in the ranks was harsh. Men slept in barracks over their horses' stalls, with the smell of urine seeping up through the floorboards. Much of peacetime routine was spent mucking out stables, combing glossy coats and sponging nostrils. The rest was all about polishing buttons and accoutrements. One sergeant complained that at the cavalry training school at Maidstone, seeing your face reflected in your scabbard was thought to be more important than the use of the sword itself. Nevertheless, boredom and drink were serious problems. As Wellington had said, some of the enlisted men were the 'sweepings of the gaols', incorrigible rogues who got used to floggings with the cat of nine tails in front of the assembled regiment. Private White of the 7th Hussars died in 1846 from the whipping he had received and months earlier, Lord Cardigan had been heavily criticised in the press for ordering 200 lashes on a man. It was not the number that bothered people, but the fact that the punishment was carried out on Palm Sunday.

For the Other Ranks, the tall tales of the 'bringer', the recruiting sergeant, evaporated as quickly as the promised bounty – more money than most men had ever seen in one place in their lives. The handsome sum disappeared in 'expenses' – riding lessons, coal and even essentials for the horses – brushes, nosebags, saddle cloth, saddle bags and sponges.

In the cavalry, the horses came first and men from the ranks were expected to care for officers' chargers as well as their own. Private Kilbourne of the 5th Dragoon Guards had to share his time between his own mount and The Earl, the horse of Lieutenant Temple Godman in his troop.

Whereas officers of the cavalry had detailed dress regulations from the 1820s, updated periodically, and bought their uniforms and equipment privately, the Other Ranks had no Dress Regulations beyond the basics laid down by the War Office. So, too, with the horses. The saddle, which was expected to last for fourteen years, was a variant of the hunting type, with a wooden tree overlaid with pigskin. The girth and surcingle that held it in place on the horse's back also had a fourteen-year lifespan; crupper and stirrup leathers would have to last for nine. Stirrups themselves, of regulation pattern steel, were not replaced until they had seen twenty years' service. Valentine Baker, who replaced Nolan as *the* cavalry expert after the Crimea, wrote, 'Ordinary saddles never get out of order and are more easily cleaned, and why a dragoon's should be such a complication of moveable straps and buckles, pilches etc. all made ... like a Chinese puzzle, has ever been beyond my comprehension.'

When the British army embarked for the Crimea in the summer of 1854, it took its horses with it. Louis Nolan had been sent by the War Department to scout ahead in Turkey to find suitable mounts, but he was to be disappointed. For all the Turks rode Arabs, there were far too few good horses to be of any use. Accordingly, the cavalry slung its mounts in canvas hammocks below the decks of their troopships. The temperatures were appalling and the animals, roped together in the dark, panicked in high seas and lashed out with their hoofs, risking injury to themselves, other horses and the safety of the wooden-hulled ships. 'No air,' wrote Colonel Hodge at the end of June. 'The horse hold and the men's decks like ovens. 82° [Fahrenheit] in my cabin.' Hodge had sixty horses on board his ship, the *Deva*, and he worried that too many of them were too old for active service. His heavy cavalrymen were *too* heavy, too, at thirteen or fourteen stone.

Fanny Duberly, whose husband Henry was paymaster of the 8th Hussars, was distraught when her favourite pony, a grey, died on board ship and had to be buried at sea. On the *Deva* one horse had to be put down. 'We ... threw him overboard after knocking him on the head,' Hodge wrote. 'It was a fine and good horse.' The worst incident of the journey may have been indirectly to do with the horses. A fire broke out in the hay stores of the *Europa*, carrying the Inniskilling Dragoons and it sank 200 miles off Plymouth on the night of 31 May. All fifty-seven horses on board went down with it, along with a number of men and one woman, a regimental wife.

In the Crimea itself, the inactivity of the cavalry became a standard joke. Lord Lucan, commanding both Heavy and Light

Brigades, was nicknamed Look-On because he merely watched the infantry action. Lord Raglan, the 70-year-old commander-in-chief, wanted to keep his cavalry 'in a band box' because he had so few of them. The Heavy Brigade was left behind at Varna in Turkey, badly hit by cholera, while the Light Brigade took no part in the battle of the Alma (20 September) at all.

Cardigan, desperate to see his first action, chased the Cossacks for seventeen days, exhausting the men and horses of the 8th Hussars and the 13th Light Dragoons in what became known as the 'soreback' reconnaissance. Five horses died and another seventy-five were unrideable for weeks.

By the middle of September, rain was causing the cavalry problems. Hodge wrote, 'The poor brutes had scraped away all the sand and were all standing in canals of muddy water, their coats staring, themselves shivering and looking only fit for dogs.' For much of the autumn and early winter, the cavalry were sleeping in the open with useless canvas tents for the men. Huts would not arrive until January. The regiments turned out in the early hours, horses saddled and bridled and would often just stand around, waiting for orders. It was hopelessly frustrating.

Then, on 25 October, action at last! A massive force of Russians under General Liprandi attacked a string of redoubts (forts) held by the Turks on the Causeway Heights. Raglan issued immediate orders but the infantry divisions were slow to move and the only unit near enough to stop the advance was the 93rd Highlanders under General Colin Campbell. Immortalised as the 'thin red streak' (later 'line') by the war correspondent William Russell of *The Times*, Campbell's men repulsed the Russian cavalry who melted away with heavy casualties.

Next in line were the Heavy Brigade, led by General James Scarlett. Like nearly every other senior officer in the Crimea, Scarlett had never seen action in his life. Nevertheless, he led his five dragoon regiments admirably, charging right through the Russian cavalry facing him. Wearing his brass helmet, Scarlett came through without a scratch but his ADC, Alexander Elliot, was less lucky. The plumed cocked hat of a staff officer was no protection at all and Elliot was hit by half a dozen sword cuts, one all but slicing off his nose. Every man of the Heavy Brigade wore their stirrup leathers at double thickness to prevent their being cut by infantry bayonets – a man without stirrups (with all due respect to the cavalry of the ancient world) was all but useless.

Colonel Clarke of the Scots Greys was riding Sultan that day, a skittish animal that could never get used to the sound of gunfire. The colonel's bearskin was lopped off and he rode the charge bareheaded. The Greys in particular rode big horses and loomed over the Cossacks on their rugged little ponies of the Steppes.

Temple Godman was delighted with the behaviour of his horse. He wrote to his father the next day:

> I rode the Earl, an excellent mount for anything of the kind, he has such pluck he will go anywhere, while I can easily manage him with one hand. He is much faster than the Russian horses; two or three times I slacked my hand and in about three strides he ran alongside any of the followers going as hard as they could.

The Heavy Brigade had routed a Russian force three times their number, for the loss of seventy-eight killed and wounded. 'The most glorious thing I ever saw,' a French general said. Even Raglan, not known to over-sing any man's praises, sent a galloper to Scarlett with the message 'Well done!'

The Light Brigade was still awaiting orders at the entrance to the North Valley and had seen Scarlett's action first hand. Captain

Morris had begged Cardigan to follow up the attack and turn the Russian retreat into a rout but the stubborn and inexperienced brigadier would have none of it. 'My God,' Morris slapped his boot with his sword, 'What a chance we are losing.' Ronald was not the only named horse we know about. Most officers in the Crimea had at least two mounts (Raglan and Hodge had five each) and although the men's horses were only known by a regimental number branded on their hindquarters, it is very likely that the soldiers had pet names for them too. George Paget of the 4th Lights rode Exquisite. Major Rodolph de Salis of the 8th Hussars had Drummer Boy and in later years, rode the animal with his Crimean Medal on a ribbon around the horse's neck. Captain Soames Jenyns of the 13th Lights rode Moses, one of several animals that had to be shot after the charge. William Morris's horse was Old Treasurer, but he may have had a second, Spitfire, with him in the Crimea.

The only man able to stop the disaster that was about to unfold was Louis Nolan. Was that why he spurred forward, trying to turn the brigade?

Although the Heavy Brigade did indeed, as Morris had suggested, support the Lights down the 'Valley of Death', Lucan recalled them halfway down, realising that half his cavalry were riding to their own destruction. William Russell, watching with horror from the Sapouné Heights, recorded the sounds of the Light Brigade charge in his account – the 'ping' of bullets, the 'sighing', 'humming' and 'whang' of round-shot, the 'harsh whir' of bursting shells and the ghastly 'slosh' as man or horse was hit. Heavy Brigade survivors told with horror the sights of the shattered Light Brigade turning back, having leapt the Russian guns. 'It was a fearful sight,' remembered Colonel John Yorke of the Royal Dragoons, 'and the appearance of all who retired was as if they had passed through a heavy shower of blood, positively dripping and saturated and shattered arms blowing back like empty sleeves as the poor fellows ran to the rear ... another moment, my horse was shot in the right flank.' Yorke himself lost his leg to canister shot.

Ronald was the first horse to reach the guns, leaping over cannon like a five-barred gate at home on Cardigan's Deene Park estate in Northamptonshire. William Morris ran his sword home to the hilt in the body of a Russian officer, then found, to his horror, he could not get it out and went down in the melee with a lance point skewering his skull. Thanks to sheer guts, a lot of luck and the skill of surgeon James Mouat of the Inniskillings, Morris lived to tell the tale.

George Paget had ridden Exquisite all the way down the valley, giving Cardigan his best support. The cigar he had lit just before the charge was a dribbling stub by the time he wheeled his horse around to try to form up his regiment. Riderless horses, panicked and exhausted, crowded against his legs, threatening to crush them and Paget had to hack at them with his sabre to prevent disaster.

'It was a mad-brained trick,' Cardigan said to his shattered brigade as they formed up afterwards, 'but it was no fault of mine.'

'Never mind, my lord,' someone shouted. 'We are ready to go again.'

'No, no,' Cardigan smiled. 'You have done enough.'

There would be no more major cavalry actions in the Crimea – the battle of Inkerman, like the Alma, was an infantry slog. It was the terrible winter that exposed the inadequacies of the commissariat in the British army. A hurricane in November had destroyed tents and had blown away equipment – even horses were bowled over by the wind. The 7-mile road from the British camp to Balaclava Harbour became a sheet of ice by December and the cavalry were expected to carry the wounded and frostbite cases to the ships bound for the base hospital at Scutari, seven days' sailing away. Many horses broke their legs and had to be destroyed.

At home, the government of Lord Aberdeen fell and his replacement, Lord Palmerston, set up a board of enquiry spearheaded by the radical MP John Roebuck. The McNeill-Tulloch Committee interviewed officers from all branches of the service and came to some gloomy conclusions. The condition of the horses, appalling though it was, was way down everybody's priority list. Lord Raglan succumbed to cholera and was replaced by General James Simpson. The Committee came to be known as the 'Whitewashing Board' – no heads rolled – and an increasingly literate and politically aware public were left wondering what it was all about. The bottom line was that the British effort had been poor, but the Russians' so much worse.

Ill with a urinary tract infection, James Brudenell had gone home in November 1854, and he took Ronald with him. Man and horse were treated as heroes, a thrilled public trying to pull hairs from Ronald's tail as souvenirs. Fourteen years later, Cardigan went out riding as he usually did, alone on a damp morning. He never came back. His horse had thrown him and the rumour was that it was Ronald, tired of the old general at last. In fact, it was a much younger, less reliable animal. Today, Ronald's head is stuffed in a glass case at the Brudenell home at Deene Park, his hoofs mounted on display.

Chapter 14

Traveller v. Cincinnati

He could make out the regiment's guidon shifting in the night breeze and knew its lettering even in the half light – Company A, 43rd Virginia Cavalry. He was Private John Alexander and he was seventeen. His left hand gripped the reins of his sorrel, his right hovered near the butt of his Spiller and Burr revolver. He carried no sabre; none of his comrades did. Swords were all right for the parade ground and the sons of bitches from West Point. He was a soldier of the Confederacy; his butternut shell jacket and kepi reminded him of that fact every day of his life.

To his right and left, the company stretched out, each man with his own thoughts, ready to pat his horse's neck and keep it quiet. Somewhere in the darkness ahead the blue-bellies were waiting for them; in Alexander's mind, with their rifles primed and ready, aimed at his head. He felt the iron lump of fear in his throat. It was not supposed to be like this. When he'd enlisted four months ago, they had told him all about the cavalry; the plumed hats, the gay ladies fawning over him, the fine, snorting horses and the thunder of the charge. At 17, he was too young to understand the odd look in the eyes of the recruiting sergeant – the look that said the old bastard was lying through his teeth.

There was a noise ahead. A snapping twig? The click of a rifle hammer? He grabbed his gun.

'Pshaw!' a kindly voice hissed in his ear. 'We're just getting in hearing of 'em. Don't be scared.'

That was Sergeant Topham, craggy, bearded, pipe-smoking, a veteran of more scraps than Alexander had had hot dinners. And he hadn't had one of *those* in a long time. The sound of the horses' hoofs crunching on the grass sounded to the boy like a funeral march – his.

'There they are!' His voice was a screech, despite the orders to keep silent.

'They're our scouts,' Topham growled, knocking the lad's revolver muzzle down. 'What the devil did you leave your mammy for?'

Afraid he might blow his own leg off, Alexander holstered the pistol and jumped at the call of a whippoorwill somewhere in the Fairfax County blackness. He tried to concentrate, clear his head of the cold terror that gripped him, stop the pounding of the blood in his temples.

There was no command for what followed; no bugle call, no trumpet blast, just the creak of leather as the whole company dismounted, slipping out of their saddles and passing their reins to the rough-riders. Alexander let annoyance take the place of terror. If he'd wanted to fight on foot, he'd have joined the son-of-a-bitch infantry!

'Charge!' somebody yelled and the lines dashed forward, firing carbines and pistols at the silent tents, chopping confused and half-asleep sentries with rifle butts and fists. All hell exploded around Alexander, the enemy's panicking horses breaking free from their roped lines and whinnying in terror. Guns exploded everywhere, powder blackening faces and blood spattering over blankets and canvas.

Alexander was running forward when something caught him across the forehead, sending him hurtling into the melee. He felt sick with the pain and rolled over, expecting to see a Billy Yank looming over him with his bayonet aimed at his heart. Instead, he saw an oak tree, minding its own business and he realised he had collided with one of its branches. As he got to his feet, the shouting lessened and the shots became sporadic. He saw black heads bobbing ahead of him as the Yankees ran. And he hadn't even fired his gun! Time he did something about that. He levelled the weapon and cocked it. There. On the left. He'd take one down for the glory of the company. The blast jolted his arm backwards and he steadied himself. It was over. No sound now except for the groans and whimpers of the wounded and the dying.

A horseman cantered through the night, ducking under the trees that fringed the Federal camp. It was the colonel himself, upright in the saddle as always, like a ramrod in the breech. Colonel Mosby, the 'Grey Ghost of the Confederacy'. 'One casualty,' he shouted to the ragged line of triumphant cavalrymen in front of him, 'but it was his own fault in chasing after a beaten enemy. Some damned greenhorn among you bastards let off a shot and hit him in the heel – but he'll live.'

Private John H. Alexander of Company A, the 43rd Virginia Cavalry, did not say a word.

The war between the States was a turning point in American history and historians are still divided over its root cause. Some attribute it to slavery, the 'peculiar institution' which had once been widespread in all thirteen of the original British colonies, but which, by 1861, was confined to the South alone. Others claim that it was all about State rights, that the federal system established after the War of Independence gave each State sovereign power to decide issues for itself and not to be at the dictate of the national government in Washington.

The war left scars, as all wars do, not least because one-fifth of the soldiers who fought never, unlike Johnny of the folk song, 'came marching home' again. Because of the still rigid tactics learned at

West Point coinciding with the ever deadlier, rapid firepower of the guns, the casualties were horrendous and more Americans died during those four years than in all America's wars before and since.

It was the election of Abraham Lincoln in 1860 that led to South Carolina and, later, other States, to secede from the Union. Lincoln was a known abolitionist and thousands of 'good ole boys' from the South could not accept that their way of life was about to come to an end. The renegade States called themselves the Confederacy, with their own president, ex-cavalryman Senator Jefferson Davis of Mississippi, their own currency and their own army and navy. The first clash of arms came at Fort Sumpter in Carolina and thousands of eager young men, North and South, rushed to the colours.

By European standards, the armies of both sides were hopelessly amateur. As had happened in Britain in the previous century, any gentleman with cash and a sense of honour could raise a regiment of horse and be made its colonel. By definition, other than the small, scattered regular army and the officer cadets at West Point, no one had any experience of war at all. It was a cosmopolitan effort – Europeans from every nation fought in the war, some emigrating just for a chance to fight. Hundreds of Irishmen, for example, still suffering in the aftermath of the potato famine, ended up in New York and joined militia units there. Eventually, even black soldiers were allowed to enlist, like the 54th Massachusetts Infantry or the 'Buffalo Soldiers' of the 9th and 10th US Cavalry. The Confederacy even set up a Native American unit, the Cherokee Mounted Rifles, whose colonel, Stand Watie, became a full general by the end of the war.

Where possible, American cavalry units emulated their European counterparts. Photographs of cavalrymen resting in camps and bivouacs show curved sabres based on the French pattern and peaked kepis, also of French design. What could sound more British than the Confederacy's Sussex Light Dragoons or the 6th Pennsylvania Cavalry, known as Rush's Lancers after its colonel?

There was a marked difference between the cavalry of the South, bred to riding as most of the enlisted men were and the North, where the roughnecks of Chicago and Hell's Kitchen, New York, barely knew one end of a horse from another. There was an East–West division too, the men west of the Mississippi far more used to the saddle than their eastern counterparts. That said, both sides rode Quarter Horses, Appaloosas, the new Standardbred and Saddlebred, Palominos and Pintos, as well as the occasional heavy draught Morgan and Mustang and, by all accounts, neither side treated its mounts very well.

As a still new nation without all the prejudices of the old, many young Americans went to war with a romantic notion of honour, glory and adventure. Every major war produces the same spirit – until the reality of the battlefield and the monotony of camp life kicks in. The war produced early cavalry successes under *beau sabreurs* like 'Jeb' Stuart. His 'Black Cavalry' wore the grey jackets of the South with black braiding and broad, plumed hats; they would not, apart from their firearms, have been out of place on a battlefield of the *English* Civil War. The reality of war meant that Rush's Lancers had abandoned their cumbersome nine-foot weapons by May 1863 and relied on carbines instead. Likewise, although they were often photographed with them, few cavalrymen actually used their sabres in action, finding greater safety in their guns.

There was a great deal of bad feeling in both armies between cavalry and infantry. Ironically, the first casualty of the war was a cavalryman, killed in a skirmish at Fairfax Court House, Virginia on 1 June 1861. The strategy and tactics adopted by both sides meant that their cavalry was used for guerrilla raids deep in enemy territory more than on conventional battlefields. This gave the cavalry the reputation of plunderers and bandits, stealing anything moveable from defenceless civilians caught up in the conflict. As one infantryman put it – 'Whoever saw a dead cavalryman?'

The lack of expertise on both sides was telling. When war broke out, seventeen of the twenty-five officers of the 2nd US Cavalry resigned their commissions to fight for the South. Most of these men had only carried out picket duty or engaged renegade 'Indians' in the far west; everything else was textbook stuff. And the textbooks were hopelessly outdated; *The System of Cavalry Tactics* was twenty years old in 1861, based on the lessons learned against Napoleon in a war the Americans had not fought.

The Federal government underestimated the need for cavalry and no new units were raised after Fort Sumpter fell. Horses were expensive. General Irvin McDowell marched into Virginia in the opening weeks of the war with 37,000 men, but only 700 of them were horsemen. 'Jeb' Stuart's Black Horse unit literally ran rings around them. Governments of all nationalities and all periods fret about the cost of war, cutting corners where they can. A regiment of horse cost nearly half a million dollars to raise in 1861 and the elderly generals from West Point shared Lincoln's cabinet's disapproval of that.

Training was also a problem. A cavalry officer took three years to train; a trooper nearly as long. As in all wars, everybody assumed it would all be over by Christmas, so was there really any point in going to new lengths? The North's defeat at Manassas (Bull Run) in June led to a drastic re-think and by the end of the year there were eighty-two mounted volunteer regiments amounting to over 90,000 men.

General Philip Sheridan, whose own West Point career was, to say the least, chequered, maintained that the ideal cavalryman should be between 18 and 22 years old and should weigh 130lb. Many Southerners provided their own horses while the North found whatever nags they could. Unlike European armies, where the new revolver was a dirty word, every Federal cavalryman was equipped with a six-shot cylinder weapon, a sabre (quickly abandoned), a carbine, saddle and harness. In the first clashes, the Confederate cavalry was streets ahead of Lincoln's boys. General William Tecumseh Sherman wrote that they were 'splendid riders, first-rate shots and utterly reckless'. He also said that they were the best cavalry in the world, although there must have been many thousands of Europeans (not to mention Yankees) who disagreed with him.

But the Federal cavalry was capable of stirring action too. In October 1862, Major Charles Zagonyi, an ex-hussar of the Hungarian army, led a charge at Springfield, Illinois that scattered a much larger force of Confederate cavalry and routed their infantry. Zagonyi's orders as his men advanced were to ignore 'every ceremonious cut' and simply right cut and thrust. He even had time, in his broken English, to deliver a rallying speech to his soldiers. 'The enemy is two thousand and we are but one hundred and fifty. It is possible no man will come back.' He gave them the choice of retreating and nobody moved. '"Fremont and the Union",' he shouted, 'Charge!' Zagonyi's unit lost fifteen dead and ten taken prisoner – 'of the wounded, not one will lose a finger'.

Even so, there was no one quite like 'Jeb' Stuart. Commanding the cavalry of the Army of Northern Virginia, he carried out a lightning raid in June 1862 clashing with General George McClellan's troops – 'Friend and foe alike were soon enveloped in bellowing [*sic*] clouds of dust, through which pistol and carbine shots were seen darting to and fro like flashes of lightning' as a rebel trooper wrote. Stuart took 165 prisoners, 260 horses and mules and destroyed camp after camp. Everyone was euphoric about Stuart's ride and only later did some come to realise that the man had far exceeded General Robert E. Lee's orders and had, perhaps, got lucky.

The problem for the South was its lack of industry. At the start of the war, the Confederacy had only two steel mills, whereas the North was geared up to the manufacture of modern warfare. Stuart himself carried a Le Mat pistol with shotgun barrel, but some of his men were reliant on old flintlocks of the Independence era. Many weapons were stolen in cavalry raids or taken from the dead bodies among the Billy Yanks; others were imported from Europe, especially Britain, at the time the 'workshop of the world', the most prolific industrial nation in history. Sabres became obsolete; Colonel Mosby, the 'Grey Ghost of the Confederacy', wrote, 'the only real use I ever heard of their being put to was to hold a piece of meat over the fire for frying.'

General John Pope of the Union army reorganised his cavalry of the Army of Virginia so that it formed one cohesive unit. This ran counter to the ante-bellum notion of all arms, inherited from Napoleon, where cavalry, infantry and artillery operated together. It made for flexibility but often left the mounted arm low on numbers and ineffectual in action. Above all, the new Federal cavalry were separated from the lumbering supply wagons, carrying three days' rations tied to their saddles and relying on foraging for the rest. By the end of the war, most cavalrymen fought as dragoons, one man in four holding the horses while the other three dismounted and fought on foot. Such tactics might have worked in the Shenandoah and Wilderness campaigns, but it would kill 'General' George Custer and his command on the Little Big Horn fourteen years later.

One of the very few cavalry clashes of the war occurred on 9 June 1863 at Brandy Station when Stuart's 10,000 men took on the 8,000 of Major General Alfred Pleasanton. Stuart was caught napping by the speed of the Union advance and this was a battle of the Napoleonic type, horseman against horseman and sabre against sabre. 'In they go, faster and faster,' wrote Edward Tobe of the 1st

Maine Cavalry, 'over fences and ditches, driving the enemy a mile or more. Oh, it was grand!' All formation was lost in the melee as cavalrymen hacked at each other, left and right. Uniform colours blurred; guidons disappeared; charge after charge was delivered. Brandy Station was a draw. It was 'a glorious fight,' one Yankee wrote, 'in which the men of the North had proved themselves more than a match for the boasted Southern chivalry.'

Most of the cavalry war was all about reconnaissance, scouting and raiding. Railway junctions were a common target, so were telegraph wires – *anything* to weaken the enemy's lines of communication. At Vicksburg in 1863, Colonel Benjamin Grierson's Yankees rode 600 miles in sixteen days.

In the West, such cavalry raids took on a vicious aspect, far removed from the gentility and chivalry of West Point and Washington tea rooms. 'Bloody Bill' Anderson and William Quantrill had grim reputations as murderers and cut-throats, using the war as an excuse for wholesale banditry. When Anderson defeated an infantry unit at Centralia, Missouri, in September 1864, he had them all shot in the head, despite official orders on both sides to take such men prisoner. Later on the same day, he executed 124 men as they begged for their lives, taking scalps and hanging heads from their saddles as the Celts had done 2,000 years earlier. From such raids emerged the outlaw brothers Frank and Jesse James, perhaps the first criminals in history to blame the circumstances of their youth for their violence.

One of the scenes that has emerged with nostalgia for the defeated South was the surrender of Robert E. Lee at Appomattox Court House, Virginia in April 1865. The horse he rode that day was Traveller, a grey on which he had been photographed months before. The Lee family had a long cavalry history – his father had commanded George Washington's Light Horse in the War of Independence and Robert himself had come second in the class at

West Point. As superintendent of the officer training school in the 1850s, he had expelled the decidedly unmilitary James McNeill Whistler for an appalling examination result.

Lee opposed slavery, like a surprising number of Southern gentlemen and he opposed the secession of South Carolina. Nevertheless, feeling stronger loyalties to Virginia, he joined the Confederacy. He invaded the North twice and held off Union attempts to take Richmond, the South's capital. He humiliated General Hooker and his Army of the Potomac at Chancellorsville, despite the loss of his ablest lieutenant, 'Stonewall' Jackson. Another subordinate, James Longstreet, let him down at Gettysburg and this was about his only failing – he was too kind and generous to men who could not reach his high standards. His last words – he died in 1870, aged 63 – were reportedly, 'Strike the tent'.

Traveller was the foal of a famous American racehorse and was originally called Jeff Davis. His dam was Flora and he was born in 1857 of the Saddlebred breed. He stood 16 hands, iron grey with black point colour. Lee bought him in 1862 from one of his captains, Joseph M. Broun, when he was named Greenbrier. 'He needed neither whip nor spur,' Broun's brother remembered, and Lee himself wrote to an artist, Markie Williams:

> Such a portrait [of Traveller] would inspire a poet ... and describe his evidence of toil, hunger, thirst, heat and cold and the dangers and suffering through which he has passed ... He might even imagine his thoughts through the long night-marches and days of the battle ... But I am no artist ... and can therefore only say he is a Confederate gray.

The horse had a very steady nerve and was not rattled by gunfire. The only time he panicked was at the second Bull Run while Lee was out of the saddle, holding his bridle. Traveller hauled Lee backwards so that he fell over a tree stump, breaking the bones in both hands.

After the war, the horse lost several tail hairs to the students at Washington College, Lexington, where Lee was superintendent. The general wrote to his daughter, 'The boys were plucking out his tail and he is presenting the appearance of a plucked chicken.'

At Lee's funeral, Traveller was led behind the caisson with a black shabraque over his saddle. He stepped on a nail months later and developed tetanus. He was shot to end his suffering. His bones now lie near his master's grave at the Washington and Lee university campus. His stable door still stands with the doors open to allow his ghost to come and go.

By contrast with Lee's dazzling beginnings, Ulysses Simpson Grant came bottom in his class at West Point and was forced to resign his commission because of drunkenness on the frontier duty during the Mexican war. A failed businessman until 1861, he was broke and begged and cajoled the role of command of the 21st Illinois Volunteers when war broke out. At Forts Henry and Donelson, then at Shiloh and Vicksburg, he proved his many critics wrong by inflicting heavy defeats on the Confederacy. In November 1863, Lincoln appointed him commander of Federal forces in the West and he rescued the Chattanooga campaign from disaster. In March 1864 he was given overall command; as Lincoln put it, 'I need that man. He fights.' He hit the South hard, avoiding pointless manoeuvre and scoring victories at Spotsylvania, Cold Harbor and St Petersburg. He accepted Lee's surrender at Appomattox personally. As T. Harry Williams wrote in *Lincoln and his Generals*, 'Lee was the last of the great old-fashioned generals, Grant the first of the great moderns.'

Grant's horsemanship was exceptional. At 5 years old, he was already doing stunts and hauling lumber with horses. At West Point, his one claim to fame was that he set a horse high-jumping record that stood for fifty years. He rode ten different horses during the war years. Cincinnati was typical of the large animals he rode, at 17 hands 2 inches, the offspring of a stallion owned by William T. Sherman. Grant called Cincinnati 'the finest horse I have ever seen'. Only two other people ever rode him – one was Abraham Lincoln. Grant was badly hurt soon after the Vicksburg campaign when his horse threw him. He was in bed for a week, but the animal was not Cincinnati.

The horse was bay and very placid, seldom batting an eyelid at anything. 'Battle sounds stirred him with enthusiasm,' wrote Captain Samuel H. Beckwith, Grant's cypher operator. 'No artist could paint the beauty of this horse in the midst of action, when the curb was required to hold him back.'

Ten thousand dollars was offered for Cincinnati after the war, but Grant turned the offer down. The horse died of old age and is buried on a farm in Maryland once owned by Admiral Daniel Ammen, who was the only man, other than Grant and Lincoln, allowed to ride him.

Chapter 15

Comanche

The Indians called it the Greasy Grass, but perhaps that was a bad translation. Captain Myles Keogh could not claim to be very proficient in the Lakota language. Hell, there were those in his Wild I Company who claimed his English was pretty incomprehensible, especially when he had had a jar or two.

He would welcome a jar about now. Even his water canteen was out of reach. He had lost his straw boater, so the sun beat down on his handsome features. Many was the fair belle who carried his likeness in their lockets; hell, Captain McDougall's sister even had a lock of his hair in hers. But Myles Keogh was not thinking of the ladies just now. He was thinking of staying alive. His company had marched with Custer throughout the night, pausing just once for the rot-gut stuff the army called coffee. A big village, the Arikara scouts had said, lay somewhere beyond the North Fork with the Little Bighorn twisting and curling to Keogh's left; a thousand warriors, the white scout 'Lonesome Charlie' Reynolds had calculated – couldn't be more.

Well, Reynolds was wrong. Keogh knew that now. Now that it was too late. He was kneeling in the greasy grass of that Montana upland, his company around him dwindling by the minute, bullets and arrows hissing through the air. The guidon of the Wild I still flapped behind him and away to his left he could see Custer's flag, the swallow-tail with the crossed sabres. He could see the general too, his hair short-cropped for the campaign, his famous buckskin jacket rolled on the saddle of Vic, held by an orderly in a shallow ravine.

And that was the problem. There was no cover for Myles Keogh, the fighting Irishman, no cover for Custer or his brother Tom. As for Reno and Benteen, where they hell were they? The general had ordered a holding operation. All right, the renegades outnumbered them five to one, but it would not be for long. He had sent a galloper – John Martin, wasn't it? – to Captain Benteen. Keogh had no idea what the message said but Benteen was with the pack train, with ammunition and water. And Benteen was with Reno. And Reno would be here, surely to God, any minute.

Charlie Reynolds would know with certainty the tribes that were creeping forward through the ravines, sheltering behind rocks, lying in the tall grass. He would know the Hunkpapa, the Ogallala, the Teton and the Sans Arcs. Keogh did not know a Sioux from a Cheyenne. All he knew was that they were getting ever closer, cutting his men down with a murderous fire. Arrows bit into the ground by his leg and he spun round, bringing down a warrior with his British Bulldog. The man staggered, then pitched forward, a single feather wound into his long, braided hair. Apart from his breech-cloth, the man was naked, his body painted in bright red with a burst of dots, like some mockery of the Stars and Stripes. Keogh flipped the body over. The man was as handsome as he was but for the bloody hole in his forehead.

Myles Keogh knew the sound of panicking horses. Behind him, in the only gulch near and large enough to hold them, I Company's horses were being held by the rough-riders and the 'whoops' and shrieks of the Indians were spooking them. Three or four of them had broken free already and were galloping away from the noise, ears flat, nostrils wide, the precious water canteens and ammunition pouches bouncing from their saddles. Damn! The captain would have to do something about that.

His bugler was down already, riddled with arrows, so he would have to shout. 'Back to the horses!' he yelled to his men. 'We're joining the General!' In twos and threes, I Company scrambled to their feet. All the Indians were on foot; the sight of even twenty mounted men would put the fear of God into them. Keogh reached Comanche, grabbing his bridle and stroking his nose. 'Easy, fella,' he whispered. 'Time for a gallop, eh?'

He swung into the saddle just as a bullet smashed into his left knee, tearing through the McClellan saddle flap and thudding into Comanche's flank. Horse and man went down, Comanche kicking and screaming in the ravine, desperately trying to get up. But there was no point in that. His rider had gone, kicked out of the saddle by the impact of bullets that smashed into his chest and ribs. The darling of the ladies, the fighting Irishman of the Wild I, would not be riding Comanche again.

The period of the Plains Indian Wars which followed the Civil War is possibly the lowest in the history of the American cavalry. Perhaps because of that, Hollywood over-compensated with a series of spectacular and highly romanticised movies (see Epilogue). At the time, very few white men accepted today's sense of outrage and equality. The indigenous American was the Redskin, a dirty savage with no morals or scruples and he stood in the way of American progress. The newspaperman Horace Greeley had urged pioneers to 'go West, young man' and the wagon trains of civilians that crossed the Rockies in search of land, gold and freedom after the horrors of the Civil War, trampled all over territory that was sacred to the Indians.

West of the Arkansas was Indian Territory where the Plains tribes lived a semi-nomadic existence following the great buffalo herds that were the lifeblood of their economy. As the numbers of white settlers grew, this territory became yet another state under the federal system and the dwindling numbers of Indians ended up on reservations, a fraction of the ground they had once owned.

A handful of Indian agents, tasked with controlling the reservations, were sensible, compassionate men, but many were not. There was a deep irony that the black slaves had been freed by the Emancipation Proclamation of 1863, but there would be no such legislation for the Indians. President Andrew Jackson said in 1829:

> There [across the Mississippi] your white brothers will not trouble you; they will have no claim to the land and you can live upon it, you and all your children, as long as the grass grows and the water runs, in peace and plenty. It will be yours forever.

And that was a good, old-fashioned American lie.

To make sure that this land became white and part of America's absurd 'manifest destiny', the army had to be kept up to strength, especially the cavalry because of the vast distances involved and the ruggedness of the Western terrain. Even so, there was the usual rumble of complaints of cost from the federal government. In 1867, President Andrew Johnson added four regiments to the existing six of the cavalry. Two of them were the 9th and 10th, officered by white men, but made up of black troops in the rank and file. Contemptuous whites called them 'Niggers' and 'Brunettes'. When the Plains Indians saw them, they called them 'Buffalo Soldiers' because of their tight curly hair and their fighting spirit. One man who despised them until he saw them in action was Major Eugene A. Carr of the 5th Cavalry. When two companies of the 10th routed nearly 500 Cheyenne warriors in 1867, he wrote, 'I have never seen such superior marksmanship among soldiers, in all my military experience.' He was a decorated Civil War hero. In the same year, Benjamin H. Grierson of the 10th made it all official – the word 'coloured' was no longer to be used. 'The regiment is simply the Tenth Regiment of Cavalry, United States Army.'

A shortage of cash kicked in, however, and by 1874, the army had shrunk from 56,000 men to 27,000, whole ranks being removed from the cavalry. This meant that officers who should have been at various forts with their units were on secondment in all manner of War Department jobs. As George Custer rode to the Little Bighorn with Myles Keogh and the 7th Cavalry in June 1876, fifteen of his forty-six officers were elsewhere.

One major problem was the change from war to peace in 1865. Custer, for example, was a major general during the Civil War, but this was a temporary brevet rank and with Lee's surrender, he became a mere colonel, involving loss of pay and prestige. It is possible that very good officers resigned their commissions rather than accept this. Those who stayed on found themselves in God-forsaken forts in the middle of nowhere, the nearest railhead three or four days' ride away. Some turned to drink; others to vicious in-fort gossip and bickering that destroyed the cohesion of men supposed to be fighting a common enemy. It is unfortunate that we do not know *exactly* what happened to Custer's five companies on the Bighorn, because the most relevant testimony came from men, Reno and Benteen, who despised him.

The rank and file of the US cavalry were a motley assortment, nearly half of them recent European immigrants. Twenty per cent were Irish, twelve per cent were German. The men who rode with Custer were from both these countries, as well as Canada and Scotland. Most were desperately poor, lured, as they were in England, by the tall tales of the recruiting sergeant and the prospect of regular pay – $16 a month, reduced to $13 in 1872. When private enterprise and greed were becoming the watchwords of the United States, it is surprising that the army could recruit at all. And desertions were common. In the year 1866–7, the newly raised 7th Cavalry lost 512 men, 35 of them absconding – the 'grand bounce' as it was known – on one night. Taking 1877 as a random year, there were 924 desertions from five regiments. In the 7th at least, this stopped when Custer ordered deserters to be hanged and carried out the threat, for which he temporarily lost command of the regiment.

Army food was bad; drunkenness and violence common. For all its reputation as a crack regiment, the men of the 7th were detested in civilian towns like Deadwood and Bismarck, often clashing with the townsfolk in saloons and brothels. Quarters were cramped, the drill monotonous and furlough almost non-existent. On top of all that, the scant pay was often seriously in arrears. Perhaps, above all – although some men did not realise it until it was too late – the enemy that the cavalry were supposed to be 'taming' fought in a completely different way and were among the finest light cavalry in the world.

In 1878 General Nelson Miles saw Crow warriors along the Rosebud River:

> Their steeds were painted in the most fantastic colors and decorated with spangles, colored horsehair and hawk's feathers. They seemed as wild as their riders, racing, rearing and plunging, yet controlled by the most expert horsemanship in the world. The warriors were painted and bedecked in every conceivable way, no two alike. Their war jackets were adorned with elk teeth, silver, mother of pearl, beads, and porcupine quills of the richest design and rarest workmanship. Some wear bear-claw necklaces and human

scalplocks dangled from their spears. Their eagle-feathered war bonnets waved in the air.

Luckily for Myles, the Crow were friendly; their arch enemies, the Lakota, had been driven out of the territory by the white man's army over the past two years.

The Plains Indians, tribes like the Assiniboine, the Gros Ventures, the Lakota, Kiowa, Comanche and Cheyenne, had taken full advantage of the arrival of the horses descended from Cortes' mounts. Many of them were pintos, made all the more exotic in battle by feathers tied to manes and tails and painted symbols, usually religious, on flanks and withers. Such horses were universally called ponies by the tribes and they were generally smaller and lighter than the mounts of the US cavalry.

The ponies were essential for all kinds of work other than war. They dragged travois, poles crossed over their backs with slings for wounded or ill people, as well as family belongings. They hunted buffalo, galloping through the great herds as their riders brought the beasts down with bow and arrow. Except where they stole a McClellan saddle from the cavalry, they rode without stirrups on simple horned saddles of wood and leather. Indians formed attachments to their horses much as whites did and it was not uncommon to find the raised graves of warriors with poles decorated with the head and tail of a favourite animal. It was easy for a trained scout to find the trail of a warband or hunting party because Indian culture had no metal artefacts; the animals were unshod.

When commanders like Custer were looking for villages, they looked for travois-trails in the dust, implying the movement of women, children and the elderly. Most camps were small, a cluster of family homes, the tipis (wrongly known as wigwams in the East) made of poles covered in dressed buffalo hide. These tents could be angled to avoid the wind and could accommodate a central fire in the grim winters on the Plains. In June 1876, which should have been the height of summer, members of Custer's 7th reported three inches of snow along the Rosebud.

Contrary to the Hollywood myth of 'red savages', the Plains tribes were deeply religious. They read symbolism in the nature around them and carried out elaborate and complicated dance rituals that seemed nonsensical to the whites. Much of this was geared to the weather – Sun Dance, Rain Dance – and every Indian prayed for the return of the wandering buffalo. Individuals like Sitting Bull were the priests of their people, given to visions and able to predict the future. His omens for Custer at the Little Bighorn proved highly prophetic.

Among some tribes, the most agile horses were called buffalo-runners and they were only used for the hunt. It is possible that there were specialist warhorses too but this cannot be proved. We know some names for the animals – Sacred Dog, Elk-Dog and God-Dog indicate the loyalty of a horse. Just as cavalry in the Civil War carried out raids, so the Plains tribes travelled miles to attack others, stealing horses and gaining prestige. Like the ancient Celts, the Indians loved to boast of their exploits. Counting coup, striking an enemy with a lance-tip or even a bow-tip, was a symbol of manly courage. While the US cavalry was out to destroy a way of life, the Plains Indian was merely showing what he could do. It is for this reason that Indian accounts of actions like that at the Little Bighorn cannot be relied upon. At least four men claimed to have killed Custer and one of them, Rain-in the-Face, said that he had eaten the heart of Tom Custer, ripped from his body on the field.

The horse raid was the most common offensive undertaken, tribe against tribe. Dances pre-figured it, involving 'medicine' from Wakan Tanka, the Great Spirit, which roughly equated with the white man's God. War honours were gained by counting coup –

at the Little Bighorn, the Cheyenne Yellow Nose grabbed Custer's guidon and used it to bash soldiers for the rest of the fight. All the time, of course, the wasi'chus (white men) were trying to kill him. Just as Europeans were awarded medals for campaigns and bravery, so the Plains warriors wore war bonnets of eagle feathers, sometimes reaching the ground. War shirts were also worn, as symbolic protection. One Blackfoot example in the Berne Historical Museum dates from the 1830s. It is painted in red and brown ochre with quillwork, beads, horsehair and human hair.

The best evidence for having counted coup was the acquisition of a gun or a scalp. All white men who spent time on the Plains were fascinated by the taking of scalps, macabre though it was. It may have been done first by French trappers in the seventeenth century, but it became standard practice, especially among the Cree and Lakota. Hair was part of an Indian's soul, which is why men as well as women wore their hair long and in braids. This may also explain their obsession with Custer – 'Long Hair' as they called him. It is ironic that the general's curly blond locks earned him the nickname of 'Fanny' at West Point and by the time of the Little Bighorn, he had had it cropped short.

The weapons of the Plains warriors were the bow, the club and the lance. Only by the nineteenth century did they acquire the white man's guns and they quickly became very proficient with them, both rifles and pistols. Indian bows were nothing like, for example, the English Medieval equivalent. They were not effective at ranges of over 150 yards, but deadly at closer quarters. Bows were recurved, either of wood or horn and arrowheads originally bone or stone, replaced later with steel. The circular shields carried by some warriors were made of buffalo hide of double thickness stretched over a wooden hoop. They were kept on tripods in camps and, like the Stars and Stripes itself, must never be allowed to touch the ground.

Several US cavalrymen who fought the Indians describe a military formation and tactical deployment they had not expected to find. Boys as young as 10 were inculcated into warrior societies and had to give gifts to the men to be allowed entry. They smoked pipes as part of the initiation and could then take part in the sacred dances and singing that varied from society to society. The Crow had their Lumpwood and Fox sects and the Big Dog society, the rough equivalent, perhaps, of the Masonic orders in Europe.

Officially, women played no part in war. Their role at the Little Bighorn was to rob the dead and carry out ritual mutilation, another ancient practice that the Europeans had chosen to forget. Dr William Bell, of the British Ethnological Society, was in the Black Hills in the 1870s and made a careful, if gruesome, study of these wounds when he examined the corpse of Sergeant Frederick Wyllyams, probably killed by Roman Nose on 26 June 1876. Bell had a particular interest in Wyllyams because the man was an Old Etonian who had, bizarrely, joined the American army. Wyllyams' right arm had been hacked to the bone – this was a Cheyenne ritual. His nose had been slit – that was Arapaho. His thighs had been slashed, so had his calves, with oblique parallel gashes. The arrows still embedded in his body had different coloured flights denoting different tribes. Only the Santee Sioux cut the throats of their enemies; early French trappers called them *coupes-gorges*.

The one known exception to women not fighting comes from the Little Bighorn itself. The week before the battle, Crazy Horse's Cheyenne attacked General Crook; and a woman, Buffalo Calf Road Woman, rescued her brother Comes-in-Sight. She was at the Greasy Grass eight days later, her cheeks daubed with circles representing the rising and the setting sun. She carried – and fired – a .44 carbine Colt Dragoon pistol.

Knowing their lands like the backs of their hands and being able to live off the land gave the Plains tribes a distinct advantage over the US cavalry. Their tactics were guerrilla warfare – hit-and-run raids and vanishing into the wilderness before the nearest lumbering cavalry train could catch them. Just *finding* the Indians was a nightmare that defeated most colonels. Only Custer, Myles and George Crook had the nous and stamina to chase the tribes to a standstill. Just as it was sheer numbers rather than criminal incompetence that outplayed George Custer at the Little Bighorn, so, ultimately, the Indians were worn down by the sheer scale of the white movement West. One by one, the great chiefs surrendered or died. Among the Apache, Cochise saw the futility of it all, while his subordinate Geronimo fought to the bitter end. He-Dog of the Oglala surrendered at Fort Robinson on 6 May 1877 and placed his sacred war shirt on the shoulders of Lieutenant W.P. Clarke as a token it was all over. Most poignant of all was Joseph of the Nez Perce. 'From where the sun now stands, I will fight no more forever.'

At the Little Bighorn, the squaws drove awls into the eardrums of George Armstrong Custer because he had not listened to the warning the Cheyenne had given him. They told him then that if ever afterwards he should break the peace promise he had made and should fight the Cheyenne, Wakan Tanka surely would cause him to be killed.

Fourteen years later, on 29 December 1890, 350 men, women and children were machine-gunned by a detachment of the 7th Cavalry at the Wounded Knee Reservation. The Little Bighorn had been called the Custer Massacre. This was the Battle of Wounded Knee.

The body of Captain Myles Keogh had not been mutilated. It was naked except for his socks and around his neck was the gold Megdalia di Pro Petri Sede given to him by Pope Pius IX for his services in the Papal Guard. Perhaps the squaws left it because they assumed it was the white man's medicine. The army buried him on the hillside where he fell and crowned the wooden marker with a wreath of the greasy grass in which he died.

It is perhaps rather fitting that the man is remembered today not for his guts, his bravado, not even the Irish tune (the *Garryowen*) that he may have given his regiment, but for the survival of his horse. When Sergeant Milton DeLacey of the Wild I found Comanche, still bleeding from his wounds, he played his part in the creation of a legend.

A number of the 7th's horses were buried on the field. Probably about a hundred animals survived. They were no use to the Indians because they were used to a diet of oats and barley and could not survive on the grass of the Plains. Comanche was given his name by Keogh in September 1868 when he was fighting Indians in Kansas. An arrow hit the horse in the right hindquarter and the shaft broke. Someone who saw the incident said the animal screamed like a Comanche and Keogh immediately gave him the name. According to the 7th's official records, Comanche was born in 1862, a 15 hands bay weighing 925lb. He carried twelve wound scars. At least six men claimed to have found Comanche at the Little Bighorn; I have gone with Sergeant DeLacey because he was with Keogh's Company and would have known the horse well. He was taken to the river, washed and led to the steamer *Far West* which had been trailing Custer's command by water. When he reached Fort Lincoln he had to be carried but was nursed back to health as a national treasure.

Colonel Sturgis of the 7th decided two years later that Comanche was 'the only living representative of the bloody tragedy of the Little Big Horn' and he must not be ridden by anyone. On high days and holidays of the 7th, Comanche was led at the head of the Wild I, a black shabraque over his back and a pair of army boots reversed in the stirrups. He died in 1891, months after Wounded Knee.

Comanche's body was dissected by a naturalist, L.L. Dyche, who made a frame and stuffed the animal at the University of Kansas in St Lawrence. And there, despite many requests to have him transferred back to Montana, he stays. As author Evan S. Connell says in *Son of the Morning Star* – 'The other horses are gone and the mysterious yellow bulldog [perhaps Custer's] is gone, which means that in a sense the legend is true. Comanche alone survived.'

Chapter 16

A Gallop Through the Empire

The British Empire of the nineteenth century was the largest in history, covering thousands of square miles and was home to millions of people. All of it was won and had to be protected by the cavalry, although, as always, all arms of the service were involved. In the sixty long years of Victoria's reign, the army fought sixty-two campaigns from Canada to China, virtually all of them, by the nature of the terrain in the particular theatre of war, needing horsemen.

The late nineteenth century saw a transformation in the British army – the days of the impossibly rich, titled officer lording over the half-starved Irish recruit were going, albeit not quickly enough for some reformers. The purchase system for officers was scrapped in 1871 and training at Sandhurst and written examinations replaced it. The length of service for Other Ranks was reduced from twenty-one years to twelve, only half of that with the Colours, the rest with the reserve. The poet Rudyard Kipling spoke for the common soldier – 'the gentleman in khaki' – and Lady Butler painted epic portrayals of military glories past. Even so, there was opposition to change. The stultifying figure of the Duke of Wellington was replaced by the even more stultifying figure of the Duke of Cambridge, who remained commander-in-chief into the 1890s. In this period, only one man, William Robertson of the 16th Lancers, rose from private soldier (in 1877) to field marshal (1913).

The advance of technology gave journalists the chance to report quickly from the field – photography helped, though not at first. In various exhibitions up and down the country, it was now possible to see fighting men in action. Roger Fenton's Crimean photographs, showing men in camp posed for the camera, opened up a new world that had previously been closed. The first war correspondent was William Russell of *The Times* and after him, no commander could ignore the press altogether.

The period also saw the publication of one of those books that acquires an influence far beyond its merits. In 1877, Anna Sewell wrote *Black Beauty*, the autobiography of a horse, featuring Captain, an ex-cavalry mount doomed to die in the shafts of a London cab. The book was an immediate bestseller, with over 100,000 copies sold and translated into French, German and Italian. Rather as *Uncle Tom's Cabin* was written by a woman (Harriet Beecher Stowe) who had never seen a slave plantation, so Anna Sewell had never ridden a horse; she damaged both ankles as a small child and was an invalid for the rest of her life.

Central to the British Empire was India, the 'brightest jewel in the imperial crown'. Much of the sub-continent was acquired by diplomacy, some by conquest, but always in order to outdo Britain's imperial rival, France. The Seven Years' War (1756–63) was a 'blue water' war in which the two nations battled it out in North America, Canada, the West Indies and India. The overall result was a resounding British victory, essentially creating what would become Victoria's empire 'on which the sun never sets'.

The most formidable enemy the British faced in India were the Sikhs in the Punjab to the north. Their artillery was particularly impressive and their gunners gave commands in French on the battlefield. In a series of battles in the 1840s, the cavalry were engaged at Sobraon, Ferozeshah and Aliwal. At the last, the 16th Lancers' lance pennons were so stiff with Sikh blood that they kept them crimped in memory of that day for as long as the regiment survived intact.

Until 1859, India's army was made up of three presidency armies owned by the East India Company. At Bengal, Bombay and Madras, the units had excellent cavalry regiments, mounted on Kathiawari horses and armed with curved swords, lances and carbines. The officers were white men, wearing topees and quilted shakos to keep the burning sun off their heads. The Other Ranks were Indians, of various tribes and sects, wearing turbans (which white officers rarely mastered) and who were led by their non-commissioned officers, the havildars and risaldars. Despite the ferocity of the Sikhs, with their razor-sharp quoits and formidable guns, once defeated they became a loyal adjunct to the British army, still fighting for the 'mother country' in the Second World War.

As well as the presidencies' armies, there were Irregular Cavalry, raised by the kind of eccentric who could only find room for their ambitions in an imperial setting. Men like Walter Fane, William Hodson and Dighton Probyn were mavericks who led their cavalry to victory in the most difficult conditions.

The regular cavalry were sent out to India both before and after the Mutiny of 1857. Today's Indian historians try to claim that this was the start of the movement for independence, but actually, it was the Sepoy war, a rebellion by *some* units of the Bengal army who believed superstitious (and unfounded) rumours that the British were trying to impose Christianity on them. When the women and children of the Cawnpore garrison were massacred by the ruthless fanatics of Nana Sahib, the government sent out Henry Havelock and others to put the Mutiny down. This they did, capturing rebel leaders and blowing them apart tied to the mouths of cannon. The upshot was that 1857 was the last of 'John Company's' wars; India would now be directly governed from Whitehall – the Raj had arrived.

Regiments posted to India saw active service on a massive scale and life there was vastly different from home service. An officer could afford more creature comforts – four horses as opposed to two and a couple of servants thrown in. Even the men had unheard of luxuries, like a syce (groom) to look after their horses. The most gruelling aspect of soldiering in India was the heat. One hussar sergeant, when the dry winds of the season began, shot himself in the head rather than face another period of it. Cornet Cooke of the 3rd Bombay Light Cavalry wrote, on 8 May 1858, 'My sword was so hot yesterday, I really could hardly hold it ... our horses are quite knocked up [exhausted] and could pursue [the rebel Tantia Topi] no further. They got no water all day.'

Commanding officers were responsible for the welfare of their regiments' horses – including their purchase. Native Indian breeds – and there were many of these – were considered too small, at 13 and a half hands, and too light for cavalry work. They were consequently interbred with English thoroughbreds. The quality of animals varied enormously, as did their prices at the hands of unscrupulous dealers. From the 1850s, Cape horses from South Africa improved the situation considerably. Colonel Valentine Baker, *the* British cavalry expert of the period, wrote that the Cape was 'the beau ideal of a light cavalry horse; strong, compact, hardy and temperate and bearing any change to any climate without deteriorating.' His own horse, Punch, 'is always in good condition and is very good tempered, and though an entire [stallion] is perfectly quiet to ride, even among mares. He has accompanied me on eight different sea voyages and is a first-rate sailor.' Such horses were not indigenous to South Africa and the Capes were Arab/Barb/Thoroughbred crosses. The most common import, however, was the Waler, another Arab/Thoroughbred cross, from New South Wales in Australia. Baker wrote, 'Their characteristics are large, ugly heads, ewe necks, good shoulders and quarters but rather long-backed, badly ribbed up. They are generally fiery and timid ...' and he clearly did not rate them as cavalry horses on campaign. Arabs were the other import and, as ever, they impressed everybody. When Baker's own regiment, the 10th Hussars, sailed from India to the Crimea in 1855, Fanny Duberly, a passionate horsewoman, fell in love with their horses straight away. Even Edward Hodge, Colonel of the 4th Dragoon Guards, who did not take to horses or people easily, wrote of the 12th Lancers' greys, that they were 'quiet, useful little animals'. Since Hodge stood at 5ft 3in, he cannot have found the Arabs all that little.

When the 9th Lancers left India in 1860, they left their horses to be bred with others at a new stud in the Punjab run by Dighton Probyn. The largest horses in India were the Turkomans, at 17 hands, but at £500 a head, only the most well-heeled officers could afford them.

Officers and men alike were astonished by the caste system in India. Winston Churchill, with the 4th Hussars in 1896, wrote 'Princes could live no better than we' and riding was essential because of the distances between cantonments and the nearest town. After the Mutiny, other than the endlessly troublesome North-West Frontier leading to Afghanistan, boredom and routine was the Indian cavalry's way of life. Horse racing, tiger-shoots and tent-pegging occupied the time, the last a useful tactic on the battlefield for skewering an opponent on foot. The cavalry game par excellence, however, was polo, taken up from a native sport by the 10th Hussars in 1869 and first played at Aldershot a year later. A local paper described the rules and commented, 'Though general remarks make it evident the new game is one most fitted for cavalry soldiers, it was more remarkable for the language used by the players than anything else.' Churchill called it 'the serious purpose in life'; Captain Robert Baden-Powell of the 13th Hussars found it excellent training for a cavalry officer. Accidents and deaths were frequent, however, to the extent that General Roberts – 'Bobs Bahadur' – insisted that helmets and chinstraps should be worn. Roberts rode a grey called Volonel, which became nearly as famous as he did. An artilleryman by instinct, he won the VC during the Mutiny for saving the life of an Indian trooper. As commander of the Punjab Frontier Force (1878–80) he fought the Second Afghan War almost single-handed, relieving the beleaguered garrison at Kandahar and achieving what umpteen armies before and since had failed to do, defeating the Afghans in their own homeland. 'Bobs' was Kipling's general – the poet of empire wrote about him twice – easily the most popular of Victoria's senior officers and went on to become the last commander-in-chief before the post was abolished in 1904.

Various historians have claimed that Britain clung so fiercely to its nineteenth-century empire because it had already lost one in the eighteenth – the thirteen colonies in North America. However the 'founding fathers' tried to dress up the creation of the United States, with pseudo-legal and pseudo-philosophical arguments (mostly pinched from the British), the War of Independence was a sordid squabble over trade and the refusal by colonists to do what every Englishman was expected to do – pay taxes.

Britain did not take the minutemen seriously until it was too late. A professional army with the reputation of the British would, it was believed, easily overawe a ragtag rabble of hunters and farmers. The problem was that because they did not understand the conventional rules of engagement, the Americans used guerrilla tactics for which the Redcoats were unprepared. While George Washington, as the only general of any standing the colonists had, tried to shape his men into European troops, teaching them Prussian drill, for example, he failed and – some would argue – the American army never did reach the standards of their European counterparts.

The War of Independence was not a cavalry war. Much of the countryside over which it was fought was mountainous, with fast-flowing rivers and few, if any, roads. The infantry could slog it out, but wide cavalry sweeps and impact charges on horseback were rare. The small group of empire loyalists, not prepared to betray their king and country, formed the British Legion under Banastre Tarleton. Some of their ranks came from the 16th and 17th Light Dragoons, the only two regiments of cavalry that the British sent out. Tarleton's Light Dragoons were equipped with short jackets and were armed with carbines and sabres. Their distinctive headgear was the Tarleton helmet, a peaked jockey cap with a huge bearskin crest that various regiments were still wearing in the 1790s.

Tarleton's men, acting as hussars, covered 105 miles over difficult terrain in May 1780, in fifty-four hours, shattering several companies of Colonel Buford's infantry who were on the way to relieve the siege of Charleston. He beat General Horatio Gates at Camden and General Sumter – the 'Carolina Gamecock' – at Fishing Creek, earning him the wholly predictable nickname 'Bloody Tarleton'.

From the colonists' point of view, Washington wrote to Congress at the end of 1776, 'From the experience I have in this campaign of the utility of Horse, I am convinced there is no carrying on the war without them and I would therefore recommend the establishment of one or more corps.' Three thousand light horsemen were planned by the infant American government, but in reality, the number never rose above one thousand. Inevitably, all ranks, from colonel to private, were amateurs – Bland's Virginian Light Horse, Sheldon's Dragoons, Baylor's and Moylan's. They all rode a mixture of Quarter Horses and the occasional mustang and equipped themselves like the British but with dark green tunics. Weaponry was very varied, including the 'long rifles' (muskets) that were totally unsuitable for cavalry use, spears and tomahawks obtained from the Indian tribes who scouted for them. As France and Holland joined the war, in revenge for losing American territory to Britain in the first place, the

Continental army (as the colonists were confusingly called) became better armed. Cowpens was an unusual American victory in 1781 when Moylan's Dragoons and McCall's Mounted Militia defeated Tarleton's Legion.

What won the war for the colonists was not their firepower, nor their determination, nor the cause for which they fought, but the fact that British generalship was average (the best commander, 'Gentleman Johnny' Burgoyne surrendered at Saratoga in 1777) and insufficient troops were sent to put the rebellion down. French troops and Dutch money at Yorktown meant that the British were outnumbered and General Cornwallis had little choice but to surrender.

If, from the British point of view, the early years of empire were blighted by the loss of America, the later years were severely shaken by the Boer War of 1899–1902. The Cape, at Africa's southern tip, had been settled first by the Portuguese, then the Dutch who found themselves at war with the indigenous tribes, the most formidable of which were the Zulu. British involvement began in earnest in 1806 as part of the ongoing conflict with Napoleon. The Emperor's own interests lay in Europe, but outposts and colonies were involved too and the Cape had to be secured to maintain the long sea route to British India.

The Boers, as the Dutch settlers were called, resented British involvement. They were both newcomers and both ignored the rights of the original inhabitants. In 1836, the Boers moved inland to Natal in the Great Trek. In essence, the British followed them. The first Boer war, in 1881, centred on Majuba Hill, where the British were routed by the superior marksmanship of the Boers who had grabbed the high ground first.

The 'great Boer war' of 1899 was another attempt by the Boers to free themselves of the British. Under their hero, Piet Cronje, 40,000 farmers organised themselves into fighting units. Various clashes usually resulted in British victories, as at Elandslaagt in October when the 9th Lancers routed the Boers with a devastating charge. The locals were reinforced by German money and their artillery was impressive. Ladysmith was besieged, along with Mafeking and Kimberley while 47,000 men formed the First Army Corps and were sent out to reinforce the original British troops. Unfortunately, their commander was Redvers Buller, grossly over-estimated as a commander and the first week of December came to be known as 'black week' with three British defeats in seven days. Buller lost men and guns in large numbers and the British defeat at Spion Kop became a dark stain on all concerned, with 300 dead, 1,000 wounded and 200 prisoners. In a phrase that would become all too common in 1914–18, 'ground gained, nil'.

The arrival of 'Bobs', now Lord Roberts, brought fresh impetus to the war. His cavalry outflanked the Boers and relieved Kimberley in the middle of February 1900. Mafeking was relieved in May, making Baden-Powell, its garrison commander, a national hero.

If things had ended there, the British reputation as a fighting force second to none might have been salvaged, but the old Boer commanders had gone, replaced by newer, more dangerous men, like Louis Botha and Jan Christian Smuts. They fought an insidious guerrilla war, of the kind that conventional armies cannot handle, using their superior knowledge of the countryside. Roberts had been authorised to burn the farms of the Boers as early as June and his replacement in November, Herbert Kitchener, took to this kind of punishment with gusto. By the time the war ended, 26,000 women and children had died in the camps set up by the British, the first of a kind that would become all too common in Nazi Germany forty years later. When the Boers finally surrendered near Pretoria, 8,000 British soldiers had been killed and 13,000 had died from disease.

Four thousand Boer soldiers had died, apart from those in the camps. The financial cost was a staggering £220 million and the cavalry had lost half a million horses.

In terms of cavalry action, one of the most enduring episodes in imperial history has to be the charge of the 21st Lancers at Omdurman in 1898. Often referred to as the last cavalry charge in history (not *quite* correct) it became famous, among other reasons, for the fact that Winston Churchill rode with the 21st that day. The regiment was the newest on the cavalry strength, raised as the 21st Hussars and because it had seen no action, wags claimed that its motto should be 'Thou Shalt Not Kill'.

The Sudan was a notoriously dangerous part of Africa, recently explored by Samuel Baker (brother of the cavalry expert) among others. General Charles Gordon had been killed at Khartoum, the capital, in 1885 and the campaign of thirteen years later was a rather belated attempt at revenge. An interesting development in terms of rapidly changing technology came from an officer of the 21st in Cairo at the beginning of the campaign – 'The tram cars at night, ablaze with electric lights and making a hideous noise with their electric bells, frighten our horses.' One reason that Wellington had disliked trains fifty years earlier was that a troop of the 17th Lancers had been spooked by the whistle and snort of a locomotive, resulting in several riders being thrown. Colonel Rowland Martin, the 21st's commanding officer, prepared his men for a desert campaign. He had 32 Australian Walers, 32 Hungarians (probably Furiosos) and 32 Arabs. Each mount carried, apart from rider and harness, hay and water necessary for five days' march. The Arabs and Walers did well; the Hungarians less so.

The men wore khaki, the dust-coloured tunics first used in the Indian Mutiny, their cork helmets covered with long neck cloths called 'lampshades'. Around their bodies dangled bandoliers of nine pouches, each one carrying fifty rounds of .303 carbine ammunition. The Lee-Enfields were carried in a leather bucket behind the rider's right leg and the swords (covered in leather or cloth to prevent the sun flashing on them and giving away their position) were fitted to the saddle on the left. The 21st had only just been made Lancers and although Martin pushed ahead with rigorous training with the weapon, his men were not exactly confident when they faced the 'fuzzy-wuzzies' and Dervishes at Omdurman. Lances had been made from bamboo, but this snapped too easily, so ash had replaced the staffs in 1885.

It was Kitchener's army, with General Gatacre as his number two. The 21st left Cairo on 31 July 1898, Churchill joining them as an attached officer (he was still with the 4th Hussars) three days later. The regiment averaged 30 miles a day and they had to ride through the full power of the sun. Fifty horses had to be destroyed, mostly because of laminitis caused by the hot sand. Private Bishop died of sunstroke; others were sent back by riverboat up the Nile. 'Dust awful,' wrote Lieutenant Arthur Pirie, the Adjutant. 'Martin never seems to think of the squadrons in the rear and had them all close up, quite unnecessary and very bad for men and horses.'

By sheer chance, a photograph taken in Cairo shows the battle formation of the regiment at Omdurman. Colonel Martin sits with his staff officers to his right and left, just behind him. Major Finn's A squadron is on the right, with Fowle's B Squadron next to him. Captain Doyle's C Squadron is next and Eadon's D Squadron on the left. The Dervishes facing them in the battle were fanatical warriors for Allah, men who had followed the 'mad Mahdi's' call to drive the white infidels out of the Sudan. They were armed with spears, swords and shields, but many had single-shot, breech-loading rifles they had stolen from the Egyptians.

Only one photograph exists from the day of battle, showing a squadron of the 21st dismounted. It was probably taken soon after eight o'clock in the morning, before orders arrived from Kitchener. The order was every bit as vague as the infamous one carried by Captain Nolan to the Light Brigade in the Crimea – 'annoy them [the enemy] as far as possible on their flank and head them off if possible from Omdurman.'

The enemy, blocking a ditch from the Nile and straggling across the Omdurman road, were nearly 1,000 strong, largely on foot and wearing the distinctive jibbah with its coloured squares to denote tribe and outfit. They had forty horses and perhaps thirty rifles. Martin saw these men and increased the regiment's pace from the walk to the trot. He had 440 men riding behind him and this attack was not impetuous or ill-conceived; he was following Kitchener's command to be as annoying as possible. What he did not know was that the Hadendoa Dervishes were supported by 2,800 Baggara tribesmen, out of sight behind the Khar Abu Sunt, a steep ditch that crossed the road.

Martin wheeled the 21st left and advanced from the south. 'Why the blazes don't we charge those bastards before they shoot us down?' Lieutenant Montmorency bellowed so that his colonel could hear him. Churchill was in no doubt that he was right – 'of course there would be a charge'.

'Right wheel into line!' was Martin's order and Trumpeter Knight sounded his bugle. All sixteen troops swung into position, Martin 30 yards ahead of his men making for the Dervish centre. *The Times* correspondent Hubert Howard was riding somewhere behind him and Captain Kenna of B Squadron was the only officer that day carrying a lance. Martin rode with no weapon in his hand – sword and revolver were still in leather. Fowle and Churchill both carried the new Mauser semi-automatic pistol and Churchill, his right arm complaining with the pain of an old polo injury, sheathed his sword and used his gun instead. At the Khar, the hidden Dervishes became apparent, screaming 'Allah il Allah!' as the 21st cheered wildly in retaliation.

Martin galloped right through the enemy but his number two, Major Cole-Wyndham was unhorsed and had to fight his way out on foot with sword and pistol. The average speed of the impact was probably 20 miles an hour, with half a ton of horse and man thudding into the Dervishes, desperately trying to skewer the cavalrymen with their short spears. In the ravine of the Khar, the enemy went for the bellies and hamstrings of the horses, slashing with their swords and knives. Dismounted men were hacked, especially around the head and chest. Martin's horse landed badly, leaping over Dervishes, but he held on and rode through them. The nine-foot lances were impressive on first impact, but in the melee, difficult to swing and use again. Many troopers dropped them and used their swords instead. These, like those in the Crimea fifty years earlier, they found too light against the broadswords of the Dervishes.

Captain Kenna was riding a grey charger called Rainbow, which landed well in the leap. He tried to save Cole-Wyndham, but the tide of battle carried them apart. Lieutenant Grenfell was thrown, his horse wounded. Before he could get up, he was hacked to death by half a dozen Dervishes. His topi was found after the battle with eleven sword cuts. Captain Eadon, by contrast, only had a dented helmet to show for the charge. Sergeant Freeman's face was badly disfigured, but he had the strength to rally his men before asking Eadon permission to fall out to 'put his nose back on'. Lieutenant Smyth's horse was hit, but he parried sword blows and dodged a spear to ride on. Lieutenant Wormald's sword bent double on impact with a mail-shirted emir, but his life was saved by a trooper spearing the man with his lance. Surgeon-Major Pinches of A Squadron had his horse killed under him, as did Private Pedder who was able to grab the reins of a riderless mount and get himself back to safety.

The fight at the Khar probably lasted less than four minutes and the 21st regrouped in the suddenly open and empty ground beyond it. Seventy-one men were unable to continue because of wounds and more than 100 horses had been killed or maimed, a quarter of the strength. Survivors reported a surreal memory of the action; several did not realise they had been wounded. Churchill wrote:

> The whole scene flickered exactly like a cinematograph picture ... I remember no sound; the whole event seemed to pass in absolute silence. The yells of the enemy, the shouts of the soldiers, the firing of many shots, the clashing of sword and spear, were unnoticed by the senses, unregistered by the brain.

As at Balaclava, many men – including Churchill – wanted to go again, to charge back through the Dervishes and rout them. Martin decided, rightly, that this would be counter-productive and that any man left behind in the Khar would be dead by now. Churchill commented drily that Martin now 'remembered for the first time that we had carbines'. Now they used them, volley fire from the saddle driving the Dervishes back, still outnumbering the cavalry four to one as they did.

Men from each squadron had the grim task of finding their comrades' bodies. Gallows humour abounded. When a private complained about his missing thumb, his sergeant said to him, 'That's nothing to make a fuss about; here's poor Sam with his head cut off and he's not saying a word.' Twenty-one bodies were recovered and buried in the level ground in front of the Khar. Captain Cordeaux oversaw the duty and broke a lance, tying it in the shape of a cross over the graves.

'Many horses were dead,' remembered Private Wade Rix of A Squadron. 'Others were trotting around riderless, some were in a distressing state with their heads down, most of them streaming with blood from the many gashes received from the fearsome double-edged swords wielded by the Dervishes.'

Gordon was avenged and when the British reached Khartoum itself they threw the Mahdi's bones into the Nile, 'which I think,' wrote Captain Sparkes of the 21st, 'rather bad form.'

Chapter 17

Warrior

We say to you, Light Horsemen, whether fields are lost or gained,
Australia's pride is safe with you, you'll keep that pride maintained;
And though they praise the French Hussars and Cossacks fierce and tall,
We know you'll prove on hard-held fields the equal of them all;
And when you ride with foamy reins and spur-raked, heaving sides,
You'll show those German Uhlan chaps the way the Bushman rides.

Jack Cox tightened his horse's girth and checked his stirrups one last time. He looked across the desert sand to the low walls of Beersheba. He remembered the name from a long time ago when he was a little boy at school in Gosport, Hampshire. He had been proud of his maps – Moses' route to the Red Sea and beyond, St Paul's missionary journey. His teacher had joked that Beersheba was not the local brewery and when you are 7, that is pretty funny.

Today had not been a joke. Little Jack Cox was now Armourer Staff Sergeant Cox of the 4th Light Horse of the Australian Imperial Force – the stripes on his shirt sleeves told the world that, along with the gusting plumes on his slouch hat. He straightened as the brigadier cantered past – the Old Boy didn't suffer fools gladly and for all Cox had served in umpteen units for the past nineteen years and seen some serious scrapping in that time, you never knew when the odd button undone or a belt that failed to shine could let you down. But the Old Boy had ridden on; the staff sergeant could relax.

The big Waler he was standing beside had been jumpy all day, what with the pounding of the guns and the rattle of machine-gun fire. The artillery had not worked; neither had the infantry. It looked increasingly like the cavalry would have to do the business after all.

Cox looked along the lines of his troop. Lads from all walks of life, bushmen from the outback, miners from Ballarat, sheep herders born to the saddle. Most of them were bronzed anyway with the Aussie sun, but the heat of the Middle East had darkened them still further and they had taken on the colour of old mahogany under their slouch hats. The Beersheba sun had bleached their khaki and produced the usual sweat stains, for all it was October.

Now, at last, the sun was beginning to dip. It must be about four o'clock and those bloody swarms of black flies had gone wherever flies go in the darkness.

'The Fourth will mount,' Cox heard the brigadier's voice, harsh, clipped, to the point.

'Bloody hell,' Jim Travis muttered in the lines next to his sergeant, 'What the bloody hell's going on?'

'We are, lad,' Cox muttered back. 'Short straw, I guess.'

The regiment mounted. And waited. There was no sound now except the occasional creak of leather. The guns were silent.

'Draw bayonets!'

Travis gulped and looked at Cox. He was only eighteen. This was going to be some baptism for him.

'Bayonets?' Travis mouthed.

Cox flicked his fingers a second before he drew his blade. 'What a day for the whole regiment to forget our swords,' he said. 'Never mind, just think of it as being two feet longer than it is and you'll be all right. And by the way ...' he hauled on his reins as the bugle blew the advance, 'a bayonet'll go through a Turk like a knife through butter. All depends on the man behind it.' And he winked at the boy.

The 4th were trotting now, the light horsemen rising in their stirrups, their bayonets at the slope against their shoulders. For all his bravado of a few moments ago, Cox knew that Travis was right. The 4th were Mounted Infantry, not cavalry. Cox had never used a sword with this outfit; the Hampshire Carabiniers where he'd started out, yes. The Imperial Yeomanry in South Africa, certainly. But not in Australia. Mounted Infantry rode to the front and while one poor bastard held the reins of four horses, the other three hit the ground and went forward with rifle and bayonet, like the poor bloody infantry they actually were.

And yet ... here they were, rising to the canter. Had the brigadier lost his mind? The heat, the shelling, the pressure of command – you heard such stories.

'Steady, C Troop!' Cox bellowed above the thud of the hoofs. 'We'll all get there.'

Ahead of them, in the Turkish trenches that ringed the town, the enemy were panicking, fumbling with their machine guns. Their field artillery, away to the right, were aiming too high, the shells whining over the heads of the light horsemen, who kept low in the saddle. At the gallop, as the bugle sounded the charge, each man was flat on his horse's neck, his sword arm extended, the sharpened points of their bayonets gleaming in the dying sun.

Not a single machine gun had opened up before the charging line hit the trenches. Cox had lost sight of the brigadier; he'd even lost sight of Private Travis. For all he'd just been urging his boys to be steady and to follow the cavalry code, his blood was up and he saw his target ahead. He rammed his spurs home and the Waler jumped the trench, Cox slashing behind him, ripping a Turk's face from fez to jaw.

A machine-gun unit was in front, the men desperately trying to load the cumbersome weapon on to the back of a braying mule. Cox was on his own, but he couldn't let *this* opportunity go. They were dragging the mule into a redoubt, a sandbagged outpost on a little hill. Once there, they could shelter and fire on the horsemen still coming up the rise.

Cox was on his own, but he realised that the Turks didn't know that. 'Come on, boys!' he yelled to no one at all. 'This way. They're in here!' He had no idea if any of the Turks spoke English. He was playing for time. He hauled on the reins and slipped the bayonet away, jerking free his pistol from his holster. Five men faced him, their hands half in the air, their faces frozen with fear. Five men. And Armourer Staff Sergeant Cox had six shots. He made a split-second decision; cavalry charges will make you do that. He fired at the man to his left and the man went down, a dark red hole in his forehead. The others jabbered at each other, their hands in the air, their mule wandering away with its deadly gun bouncing on its back.

'For gallantry in capturing a machine gun and a crew of five gunners single-handed in a redoubt south-east of Beersheba, Armourer Staff Sergeant Arthur John Cox is awarded the Distinguished Conduct Medal.'

Arthur Conan Doyle remains famous today for his Sherlock Holmes stories, but he had served with the Medical Corps in the Boer War and had a great deal to say on military matters – 'Passing on to the cavalry,' he wrote in 1903, 'we come to the branch of the service … most in need of reform. In fact, the simplest and most effective reform would be [to] abolish it altogether … lances, swords and revolvers have only one place – the museum.'

But the powers that be disagreed. Colonel Douglas Haig of the 16th Lancers wrote in 1907 – 'the role of Cavalry, far from having diminished, has increased in importance ... it alone is of use in the service of explorations and it is of capital importance in a general action.'

The cavalry training regulations for the British army, published for the government by Eyre and Spottiswode in 1904, runs to 376 pages and all cavalry officers and NCOs were expected to know it by heart. With a preface by Roberts as commander-in-chief, it deals with every aspect of cavalry life, from training horses to parade-ground drills, to saddlery, army organisation, physical fitness and horse diseases. Acknowledging the firepower of the Boers in South Africa, whole sections of the books are given over to the 'Rifle, Short, Magazine, Lee-Enfield Mark I, complete with detailed diagrams and instructions for firing'.

What does this mean? Roberts asked his readers:

> It means that instead of the firearm being an adjunct to the sword, the sword must henceforth be an adjunct to the rifle; and that cavalry soldiers must become expert rifle shots and be constantly trained to act dismounted ... I admit that occasions may arise where the charge of a cloud of horsemen ... may be of inestimable value and may change a success into a brilliant victory or convert a retreat into a rout.

So Roberts might have foreseen the charge of the light horsemen at Beersheba, but he could hardly have been ready, any more than anyone was, for the horrors of trench warfare; Kipling's 'master-gunner' died the year before Archduke Franz Ferdinand was murdered by a terrorist fanatic in Sarajevo.

At the beginning of the First World War, which, everyone was confident, would be over by Christmas, Britain had 25,000 horses, impossibly few against the organisation and preparedness of the Kaiser's Germany and the huge resources of Emperor Franz-Josef's Austria–Hungary. There was a need to provide mounts quickly in September 1914 and the Remount Department, set up in 1887 to replace the regimental provision of horses, went into overdrive. Two years before men were conscripted, horses were being requisitioned from stables and farms all over Britain and from various parts of the Empire – Arabs from India, Walers from Australia, Quarter Horses from Canada. They had to be 15 hands tall and over 3 years old, strong enough for pulling guns as well as for arduous riding.

Optimists in September 1914 looked at the geopolitics. Germany and Austria had got themselves into a lose-lose situation – a war on two fronts. The Tsar had the biggest army in the world, the Russian juggernaut and his Cossack cavalry were famous throughout Europe. The combined military efforts of Britain and France – the country which had beaten Napoleon and the country which had bred him – would easily contain the ambitions of the Central Powers from the West. Turkey was an irrelevance – the men of 1914 were the grandsons of those who had fought in the Crimea, where 'Johnny Turk' was seen as a cowardly joke.

The reality was very different. Nicholas I's Russia, desperately backward and tottering, was rotten to the core. It would be defeated every time it faced the Germans and collapsed under the weight of two revolutions in 1917, leaving Germany with a war on just one front. The Turks, stiffened by German technology and professionalism, did surprisingly well at Gallipoli, the brainchild of Winston Churchill, the ex-hussar, and led to a punishing campaign in 1915 which achieved nothing. In the West, the French had already had a bloody taste of German militarism; in the Franco-Prussian war that had created Germany in 1871, the Germans had twenty-six railway lines to take their troops quickly to the front; the French had one.

What no one had realised, in the high commands of any of the combatants, was the sheer power of heavy artillery and the damage the guns could do. Neither did they understand the rapidity of fire of modern machine guns which could destroy a regiment in minutes. The solution was to dig deep, so that a whole line of trenches faced another on either side of No Man's Land from Belgium to the Ardennes. This became the killing ground for four years and the various 'pushes' grimly followed the same routine – days of artillery bombardment to smash trenches, create shell craters and demoralise the enemy, followed by the shrill, piercing whistles that sent infantrymen 'over the top', picking their way through barbed wire to reach and overwhelm the enemy front-line trenches. The slaughter was unbelievable; on 1 July 1916, the first day of the Somme, there were 60,000 casualties (twice the number of men under Wellington at Waterloo) and no ground was gained.

As for the cavalry of both sides, the horses waited patiently behind the lines, ready for the breakthrough that would never come. The British government spent £36 million on requisitioning horses, most of them transported from the United States and Canada. The problem of shipping horses, so acute in the Crimea, had still not been solved and 3,300 animals were lost at sea. In June 1915, 1,400 horses and mules were abandoned after their transport, the SS *Armenia*, was torpedoed by a German U-boat. The most common reason for horse deaths was 'shipping fever', actually pneumonia.

By this time, partially due to the losses of animals in the Boer War, the Blue Cross offered medical help for horses and Our Dumb Friends League kept a watchful eye. In the regiments, veterinary officers were kept busy, inspecting horses every day to prevent the outbreak of mange. Until it was discovered that over-clipping of coats could lead to cold and exposure, cavalry horses were clipped every month. Fodder was a problem. In the war zone, grass ceased to exist, everybody floundering in the mud of Flanders churned up by marching feet and the impact of shells. A horse needs ten times the daily intake of a man and supplies of grain were the biggest single financial outlay of the war, even more so than high-explosive shells. Because of the British naval blockade, the German cavalry resorted to mixing sawdust with their horses' bran which led to starvation. Three-quarters of the horse deaths in the First World War were caused by disease and exhaustion, but thousands of animals were treated for bullet wounds, the effects of chlorine and mustard gas, even shell-shock, which some generals said did not exist. Injured horses were often shot by regimental farriers on the spot, using a special device that fired a .310 calibre bullet into the head; others were taken to field hospitals in specialist carts. There was never enough stabling and most horses were picketed in the open in long lines. Even so – and the weather must have been a factor here – the annual death rate during the war was 15 per cent; in the Crimea it had been nearly eighty.

As it became obvious to all concerned that no breakthrough was forthcoming, cavalrymen were hived off into more relevant units. The Yeomanry, who had been raised largely to defend Britain against Napoleon, had seen their first overseas action in the Boer War. Now, with the creation in 1908 of the Territorial Army, they fought alongside regular cavalry on all fronts. In practice, they became infantry, artillery and cyclists and never had the chance to show what they could do as horsemen.

On average, a cavalry horse, by 1916, carried twenty-two stone, depending on the weight of his rider. In theory, but rarely in practice, a horse had to travel 40 miles a day with this weight. The saddle had been redesigned for maximum comfort to man and horse in 1912; it was the Universal Pattern, still used for mounted ceremonial today. The bridle too has survived, with its Portsmouth bit designed to take a double rein attachment, at the bar and the cheek-pieces. It was made to slip free easily to allow a horse to feed and drink.

In terms of weaponry, the last pattern sword of the British cavalry was the 1908 pattern (with a fancier version in 1912 for officers). After a century of argument, the army's top brass had decided that a straight blade designed for thrusting was the ideal cavalry weapon. This was strapped to the saddle on the left side and could be drawn with one hand. The bowl guards were usually painted khaki or green to avoid flare from the sun, as in the Sudan. In a wet and boggy France, the point of this was less obvious. The 'lance lobby' was strong in Britain, as it was in France and Germany and both bamboo and ash were used. Lancer regiments had leather 'buckets' fitted to their right stirrups to support the weapon when not in action. The rifle, carried in another leather 'bucket' to the rider's right, was the short Lee-Enfield weighing about 8lb. Its muzzle velocity was 2,000 feet per second and it had a considerable kick. Ammunition for the gun was carried in bandoliers both over the rider's shoulder and around the horse's neck. Officers and NCOs carried revolvers, usually Webleys, firing six shots, in holsters fitted to their Sam Browne belts, named after the one-armed hero of the Indian Mutiny who had invented it to make drawing a sword with one hand possible.

Although a soldier wore no armour in either the British or German armies, the French at first still wore breast and back plates and dragoon helmets with horsehair plumes. The Germans adopted the functional stalhelm for all units, replacing the less effective pickelhaube of the late nineteenth century. British cavalry wore the tin hat or 'battle bowler', as it came to be known, only really effective for glancing blows and flying debris; it could not stop a bullet.

Gas masks or respirators were worn by man and horse from 1916 and soldiers had to care for their horses injured by gas by washing them and binding their eyes. A nosebag of wet hay was an improvised gas mask before real ones were developed, but the war was over before these could be issued.

As well as weapons, a warhorse of the Great War carried wallets containing horseshoes and nails, two blankets, a mess tin (strapped to the sword scabbard), a canvas bucket (for horse-feed), a cornsack, greatcoat, spare boots, groundsheet and two grenades. Each regiment had a detachment that used the Hotchkiss machine gun from 1916, although it could not be fired by one man and had to be carried on pack horses or mules.

Although the Western Front was epitomised by trench warfare, high-explosive bombardments, gas and stagnation, cavalry actions did sometimes happen and one, in particular, had strategic importance. Every spring, both sides would regroup to launch offensives which would, hopefully, bring the war to an end. The spring of 1918 was crucial, because although Russia had collapsed and made the 'robber peace' with Germany, the United States had joined the Allies and, after years of isolation, was sending troops to France.

The Germans launched an offensive in March, making for Amiens, and 'Galloper Jack' Seely was ordered to stop them. Seely was the son of a baronet, typical of men of his class who combined military with political service. He fagged for Stanley Baldwin, the future prime minister, at Harrow and graduated from Trinity College, Cambridge, before obtaining a commission in the Hampshire Carabiniers Yeomanry in 1889. He was a captain three years later and was called to the Bar in 1897. With the outbreak of the Boer War, he must have been torn between politics and the army; a captain in the newly organised Imperial Yeomanry and MP for the Isle of Wight, where his family owned land.

On 11 August 1914, a week after the declaration of the First World War, he found himself in France with Kitchener's 'contemptible little army' (as the Kaiser called them), the British Expeditionary

Force. The horse he rode then – and which carried him relatively unscathed through four years of war – was Warrior, a 15 hands 2 inches bay with a white star on his forehead, born at Yafford in the Island in 1908.

Seely had been mentioned in despatches in the Boer War and after service as a staff officer, during which he lent Warrior to the commander-in-the-field, John French. He was given command of the Canadian Cavalry Brigade with the temporary rank of brigadier in January 1915.

The Germans had occupied a ridge overlooking the Avre river near the village of Moreuil and Seely's job was to prevent the river crossing. He rode Warrior ahead to scout the position for himself:

> I saw at once that the position was desperate, if not fatal. If the enemy captured the ridge, the main line from Amiens to Paris would be definitely broken and I already knew that when that happened the two armies – French and British – would be compelled to retire; the French on Paris and our army on the Channel ports … All that we had fought for, and bled for, for nearly four years would be lost.

Seely had 1,000 cavalrymen at his back, a ragtag mixture (as all First World War units were) of former civilians caught up in the whirlwind of war. There were clerks who had barely known one end of a horse from the other; cowboys from the West eager to mix it with 'the Hun' long before their government decided to act; even Mounties in search of adventure beyond the Rockies. Officially, they were the Royal Canadian Dragoons, Lord Strathcona's Horse and the Fort Garry Horse that Seely kept in reserve. The Germans were already camped in the Moreuil woods, hacking at the undergrowth to build a line of trenches. Seely sent a squadron forward on foot, armed with carbines and Lieutenant Gordon Flowerdew of Strathcona's drew up another squadron to cut off the enemy retreat.

As he advanced, he ran into German grenadiers armed with machine guns. The lieutenant had no doubt about his next move. 'It's a charge, boys,' he shouted. 'It's a charge!' but his bugler went down before he could blow the command. The Strathconas crashed through the German lines and back again, sabring the machine gunners in the old school tradition. Flowerdew, fatally wounded, was awarded a posthumous VC.

Seely came to be known as 'the luckiest man in the army' and Warrior 'the horse the Germans could not kill'. A sniper brought down the mount alongside Warrior, but Seely's horse was unhurt, even when he galloped forward to fix a pennon for the supporting Royal Flying Corps to aim their bombs at. The action at Moreuil Wood saved Amiens and stopped the German offensive in its tracks.

Warrior survived it all – the Somme, Ypres, Passchendaele and Cambrai, more than once being buried by the flying earth of high-explosive shells. 'I have seen him,' Seely wrote later, 'even when a shell has burst within a few feet, stand still without a tremor – just turn his head and … look at the smoke of the burst.'

Rather as Napoleon's 'grognards' (grumblers) recognised 'Coco' before the Emperor, the Canadians would call 'Here comes Warrior!' before 'Here comes the general'. Warrior came home to the Mottistone Estate in the Isle of Wight where he lived to be thirty-three.

But if the Western Front did not have the scope for cavalry action, the Middle East did. Here, General Edmund Allenby, known as 'the Bull', was a force of nature. Lantern-jawed and clear-eyed, Allenby was a natural cavalry commander. He had been commissioned in the 6th Inniskilling Dragoons in 1882 and commanded a column in the Boer War. He had been colonel of the 5th Royal Irish Lancers and commanded the 4th Cavalry Brigade. Between 1910 and 1914, he was Inspector-General of Cavalry, fully aware of the increasing criticism being levelled at his area of the service.

It did nothing to help his legendary bad temper that Allenby was given no opportunity to shine until 1917 when he was transferred to Palestine (then a British protectorate) to lead the Egyptian Expeditionary Force. Rather like Napoleon in Italy in 1796, he found the units there with low morale and no real purpose. The press at home neglected them and all the money and materiel was being thrown at the Western Front, where most generals and the prime minister, David Lloyd George, believed the war would be won.

Allenby had an army of 80,000 men, a quarter of whom were cavalry. Despite his own impeccable regular army credentials, Allenby's troops were Yeomanry and the untried bushmen, Light Horse and Mounted Rifles of New Zealand and Australia. The 'Anzacs' had already been bloodied in Gallipoli against the Turks in 1915 where a badly planned campaign (largely by Churchill and Kitchener) had gone wrong. The casualties were heavy and no ground had been gained, but the colonists had proved themselves stoic fighters. The reward, for many of them, was to be sent to the trenches of Flanders from the trenches of Gallipoli.

Those sent to Palestine after a couple of weeks rest and recreation became light horsemen, operating as hussars and light dragoons had in Europe for nearly two hundred years. In the appalling heat of the desert, water was crucial and the biblical town of Beersheba had plenty of wells. If Allenby was to take Gaza and ultimately Jerusalem, he must get to those wells first. With a clever series of feints and working at night, Allenby's infantry bombarded Gaza, diverting Turkish defences there and leaving Beersheba relatively exposed.

On 31 October, soon after four o'clock in the afternoon when the power of the sun was just beginning to wane, the men of the

4th and 12th Australian Light Horse advanced across the burning sand, first at a walk, then a trot. They wore their famous slouch hats pinned up on the left side and their horses had anti-fly fringes fitted to their bridles. Only officers carried swords; everybody else used their bayonets instead. It was a textbook charge; every cavalry commander in this book would have recognised it at once, but it was only possible because Allenby's artillery had already blown apart most of the Turkish machine-gun emplacements facing them.

The light horsemen leapt the ditches, dismounted and skewered the disbelieving Turks with their bayonets before remounting and galloping hell-for-leather for the town, saving its vital wells in the process. They then, with the Warwickshire Yeomanry, chased the Turks to Gaza and beyond, driving a wedge between the Turkish forces and keeping them separate.

Allenby walked into the holy city of Jerusalem in December 1917 as a mark of respect. Unpopular and bullish he may have been, but he would not ride in as a conqueror on a white horse; Christ himself had come this way on the back of a donkey. In the next phase of the campaign, the general made sure that swords were issued to the light horsemen. 'I consider the sword,' wrote Brigadier Wilson of the 3rd Australian Light Horse, 'has a great moral effect on both the man carrying it and on the enemy.' The Anzacs, under their General Henry Chauvel, went on to achieve more miracles at Megiddo and Aleppo. As Field Marshal Archibald Wavell said later, this was 'the most striking example of the power of the cavalry arm in the whole history of the war'.

The total losses among horses between 1914 and 1918 was 256,204, compared with 541,714 for the French army.

On 15 September 1916, in the bloody fields along the Somme in Northern France, a new concept lumbered through the mud. It had been raining heavily and the loud, clanking machine dipped into the water-logged shell craters and stuck there, caterpillar tracks grinding and rear wheels spinning. It, and the forty-nine others that followed it, was the invention of a Royal Engineer officer, Lieutenant Colonel E.D. Swinton. For reasons of high security, the prototype had been stored in a large wooden box and simply labelled 'Tank'. And the noise of these apparently useless, lumbering behemoths that day on the Somme sounded the death-knell of horsed cavalry.

Chapter 18

A Gallop Through Hollywood's Horses

The nearest that most of us – probably all of us – will ever come to a cavalry charge is to see it on the big screen. The spectacle of galloping horses is such that it is lost on television – only the cinema can give it its full impact.

Clearly, other studios than those in California made movies that featured horses – any film set in the past is bound to feature them – and this chapter looks at how horses have been treated on celluloid in the West generally.

The following list is a highly selective potpourri of horse movies, chosen not at random but because each one has a relevance to the history of the warhorse.

The Old Testament, chronologically, provided film makers from the silent era to much more recent decades with a wealth of excitement. By definition, the stories chosen by the movie moguls are epic, full of high drama and action, providing, before the Hays Commission 'cleaned up' Hollywood, just the right amount of titillation to hold audiences who knew that Delilah, the Queen of Sheba and Salome were femmes fatales who slept their way to the top. Audiences expected to see some of that on the screen. Faithfully, the movies that dealt with the Hebrews, the Hittites and the Egyptians had plenty of chariot work, keeping Second Unit directors and their teams on their toes. In *The Ten Commandments* (1956, Cecil B. DeMille), the children of Israel would have stood no chance against the pursuing army of Rameses II, even if Moses was Mr Epic himself, Charlton Heston. By modern standards, the special effects of the parting of the Red Sea are almost laughable, but the chariot work was impressive. So was it in *Solomon and Sheba* (1959, King Vidor). Having been robbed of his prey as Rameses, Yul Brynner made the mistake of falling for the curves of Gina Lollobrigida. Both she and Solomon were real historical characters, but the romance between them consisted of perhaps one political meeting – they were actually rival heads of state. The reason that *Solomon and Sheba* deserves a mention here is that the cavalry, hurtling towards the enemy, is totally blinded by the highly polished shields of the opposing infantry, a sneaky tactic not seen before or since. If it were *that* simple, why did no one else in the Old Testament try it?

Fast forward to the two easily forgettable takes on the life of Alexander the Great. In a way, Alexander's career defies belief (as does Napoleon's) because he was the greatest general of his time and that sheer uniqueness creates problems. Richard Burton played the Macedonian in the eponymous film of 1956, directed by Robert Rossen. Colin Farrell, complete with blond wig and Irish accent, tried in *Alexander* (2004, Oliver Stone) to bring him to life. Neither film convinces, and in the Farrell version, Bucephalus is depicted as a black stallion, although we have no idea of his actual colour. Black – the colour of Rupert of the Rhine's mount, for example – is equated with evil and death, white with honour and goodness. All that said, any director worth his salt has the problem of no stirrups for the cavalry before the sixth century. Today, that has to be balanced against the modern obsession with health and safety and the very real problems of accident insurance. Farrell does pretty well, but a recent television series on ancient Rome went straight for the latter – it was stirrups all round, not to mention 1902 pattern bits!

And, talking of the Romans, we must mention *Spartacus* (1960, Stanley Kubrick). As a sword and sandal epic, this film is among the best, with Dalton Trumbo writing the screenplay from Howard

Fast's novel. Trumbo was a Communist, black-listed by Hollywood and that meant that the star, Kirk Douglas, and Kubrick had to tread carefully. Hollywood in the era of Senator Joe McCarthy was obsessed with the 'reds under the bed' threat from post-war Communism, to the extent that Trumbo's name appears only once in the brochure released at the film's premiere. In those heady pre-CGI days, 8,000 extras from the Spanish army mastered the movements of Roman legionaries on the same soil that real legionaries marched 2,000 years earlier. Nino Novarese was the historical adviser and got most things right, but in one significant scene, he fails. As a gladiator, Spartacus fights with a Thracian sword, which is a variant of the gladius carried by legionaries (although it should actually have had a curved blade). This was an infantry weapon, rarely more than two feet long. The cavalry (which Spartacus led in his final battle against Rome) used a three-foot version called a spatha. One of the most common stills from the film shows Douglas on his chestnut horse raising his almost useless infantry sword to signal the advance. Minutes later, he has to reach far out of the saddle to slash at the infantry around him. A brilliant film, but military historians are picky people.

And, talking of Kirk Douglas, *The Vikings* (1958, Richard Fleischer), like *Spartacus* is a rattling good yarn, this time based broadly on the Scandinavian sagas of exploration, rape and pillage eloquently recorded by the Icelander Snorri Sturluson in the thirteenth century. The Vikings were brilliant sailors and brutal marauders, but no one equates them with cavalry. The film's interest lies in the brief scene in which Einar (Douglas) rides into his village on a Fjord horse. It is a genuine Scandinavian breed, complete with dun colour and upright mane which looks very authentic and could literally have trotted out of the eighth century.

In *El Cid* (1961, Samuel Bronston), Charlton Heston rode *two* Babiecas. Filming in two locations, Spain and Italy, it was easier to find an equine double for the Cid's horse rather than ship the original. A great deal of money and talent was spent on recreating the world of El Cid, and it is a brilliant film, but in the end it was Hollywood Medieval rather than authentic eleventh-century Spain. In the excellently choreographed tournament for the throne of Calahorra, Heston and his opponent (the enigmatic Christopher Rhodes from the Isle of Wight) ride horses barded as in the thirteenth century, wear basic plate armour (ditto) and hammer each other with two-handed swords not made until the sixteenth century – and all this for a man who died in 1099! The cavalry in the film were genuine horsemen from Madrid's Municipal Honour Guard.

By sheer coincidence, Nikolay Cherkasov, playing the Russian hero Alexander Nevsky (1938, Sergei Eisenstein) does not look unlike Charlton Heston. The film's plotline centres on the invasion of the Medieval Russian state of Novgorod in the thirteenth century. The 'baddies' are the Teutonic knights and, since the movie was made at a time of sour Russo-German relations, the invading infantry wear Stalhelm-style helmets of First World War vintage. In accordance with Communist doctrine, the project was anti-Church and anti-Catholic. The music was by Sergei Prokofiev, who, for all his brilliance, was a Stalinist, as was Eisenstein. It is noticeable that when Stalin went over to the enemy and sanctioned the Molotov–Ribbentrop pact in August 1939, *Alexander Nevsky* was withdrawn from circulation.

The high spot of the action was the famous battle on the ice, when the Teutonic knights drowned as their combined weight cracked the frozen covering of Lake Preipus. All credit to Eisentstein's team that, as the film was shot in high summer, trees had to be painted blue and dusted with chalk to suggest midwinter and the cracking ice is actually melted glass. How many horses were killed and injured is unknown – the Soviet Union kept all such statistics close to its chest.

The Mongols have not fared well in the western cinema. Gary Cooper made an unlikely Italian traveller in *The Adventures of Marco Polo* (1938, Archie Mayo), charting very little of the reality of his visit to the court of Kublai Khan in the 1270s. Omar Sharif's *Genghis Khan* (1965, Henry Levin) was dreadful. Almost everyone in it was miscast, especially Sharif himself. In a different era when such things would not have occurred to any Hollywood mogul, we have a Lebanese Syrian playing the great warlord, two Englishmen (James Mason and Robert Morley) as Chinese mandarins and an Irishman (Stephen Boyd) as the baddie, Temujin's rival, Jamuga. We get little sense of the scale of the slaughter that Genghis initiated, with piles of bodies and burning villages, but it is the *horses* that are so wrong. Perhaps because the sight of real Mongols riding the little ponies of the Steppes is almost comical (the world has forgotten how terrifying they actually were) we have the usual reliance on a mix of thoroughbreds and Quarter Horses with modern harness. About the only nod in the direction of history is animal skins thrown over the saddles. The film was shot in Yugoslavia, which did not recognise animal rights and the horse falls are horrific.

'I will tell you,' says the voice-over for Mel Gibson's *Braveheart* (1995) 'of William Wallace. Historians from England will say I am a liar. But history is written by those who have hung [*sic*] heroes.' The problem with historical movies is that they inevitably take sides. For reasons of simplicity and to grab the emotions of the audience, we have to see life through a single perspective or the result is a chaotic mishmash. Gibson has form in making anti-British films. His debut in *Gallipoli* (1981, Peter Weir) is fair enough – the 1915 campaign against Turkey was an unmitigated disaster, badly planned and executed. *The Patriot* (2000, Roland Emerich) the War of Independence from a colonial viewpoint was a travesty, and this was continued in *Braveheart*. William Wallace came across as a wronged freedom fighter in the film, whereas he was actually a murderer and a horse-thief.

The cavalry work centres on the battle of Falkirk in 1298, one of the rare examples of a Scottish victory. The Scots, like their ancestors against the Romans, relied on wild charges with as much noise as possible – hence the bagpipes – although by Wallace's time, they had developed the schiltron infantry formation. Stills from the film show the English cavalry wearing unhistorical leather bards that look like butchers' aprons and their modern bits are disguised with sheepskin. What *is* impressive is the use of slow motion to emphasise the charge of the destriers. They are clearly hairy-hoofed thoroughbreds, but the speed of their advance, reduced in effect to a trot, gives the feeling of a lumbering juggernaut against which nothing can stand.

For George Macdonald Fraser in his *The Hollywood History of the World*, Laurence Olivier's *Henry V* is 'the best film ever made', which is pushing it a bit. It begins in the Globe theatre on London's South Bank and only later changes tack to become a realistic 1415, filmed largely in Ireland. The armour is good and the sight of hundreds of arrows hurtling through the air like rain to William Walton's magnificent score is truly uplifting. And this was precisely the point. The film was made as propaganda in 1944 (see *Alexander Nevsky* above) and the charge of the French knights destroyed by the Goddamns, the English archers, is excellent, *but* it did not happen in glorious sunshine! The 1989 remake by Kenneth Branagh was far more accurate, everybody floundering in the mud caused by a torrential downpour in a congested area in late October. Olivier's version is all bright heraldry and colour – the toymakers Britains produced a series of metal knights of Agincourt as an early example of the lure of film merchandise.

One of the scenes that everybody remembers from the film is the Duke of York being hoisted into his saddle by a crane. Until the 1960s it was routinely assumed that this was largely how it was for the Medieval knight; they were formidable in attack but, once unhorsed, were helpless, like, in the usual cliché of the time, 'a beetle on its back'. In fact, the heaviest armour made (in the sixteenth century) weighed less than a British squaddie's kit in the Falklands War in 1982. For the record, the Duke of York *was* crushed to death at Agincourt, but that was because of the sheer press of bodies around him and had nothing to do with his armour.

Another Shakespeare/Olivier co-production was *Richard III* (1955) and for Bosworth Field, read somewhere in Spain. Shakespeare, of course, writing for the limited scope of the Elizabethan stage, had no concept of the battle's actual events and he could not show them even if he had. Olivier's direction – helped by his fight arranger Bernard Hepton – is not a bad re-enactment. White Surrey probably wore Gothic plate armour that day while Olivier's grey only wears the same harness he wore throughout the film. Richard's charge down Ambien Hill is fairly faithful to events, although there is no record of the king grabbing somebody else's horse to carry on fighting. Incidentally, the Ian McKellen remake in 1996 had a fascinating take on the famous unhistorical 'A horse! A horse! My kingdom for a horse!' McKellen's *Richard* is set during the 1930s with the king's white boar badge transformed into a pseudo-swastika. Bosworth, filmed in the derelict Battersea Power Station, is all about heavy artillery and bombs. To get the 'horse' line in, McKellen's jeep breaks down and he tries to resort to the older form of cavalry to carry on fighting. Ingenious or corny? It's up to you to decide.

Many commentators have waxed lyrical about the meeting of the Zhaparotski Cossacks in *Taras Bulba* (1962, J. Lee Thompson), again starring Yul Brynner. The man was mythologised by Nikolai Gogol at a time when the Ukrainian cavalry of the Steppes were breaking out from Polish control and before Muscovy became all-encompassing Russia. There are far too many grey horses in the film, which is given no explanation, but the wild fervour of the Cossacks and their mounts is well handled.

On the face of it, Richard Lester's *The Three Musketeers* (1973), based on the Alexandre Dumas novel has little to do with horses, but it is included here to illustrate a sad point. During filming of the sequel, *The Return of the Musketeers* (1989), in September 1988, comedian Roy Kinnear, playing the servant Planchet, died after his horse slipped and he was thrown. Whether Kinnear insisted on doing his own stunts or whether this was an oversight on the part of the director is unknown; but it is a reminder of how dangerous horsemanship can be, even when it is carried out for entertainment purposes.

Cromwell (1970, Ken Hughes), starring Richard Harris and Alec Guinness, plays fast and loose with history and *just about* gets away with it. Harris is hopelessly miscast as the future Lord Protector and the script give him far too central a political role. He was, for example, merely an obscure MP/Captain of Horse at Edgehill in 1642. Guinness as Charles I, however, is magnificent – a dignified, unbending idiot, largely responsible for his own downfall. Two horse-based problems stand out in *Cromwell*. The first is Prince Rupert of the Rhine, played with panache by Timothy Dalton (later James Bond). Through no fault of his own, Dalton rides a grey horse (Barbarie was black) and his dog, Boye, was a hunting poodle, about the same height as an Alsatian, not the toy breed Rupert carries on his arm in the film. The second is the situation recalled by

Robert Morley in his memoirs. He was playing the Parliamentary commander the Earl of Manchester and, with his considerable bulk, sitting his horse on soft ground, began to sink slowly up to the animal's hocks in the mud. To cap it all, he realised that he had left the protective plastic covering on the plumes of his hat – with today's excellently crisp stop-motion button, you can clearly see this in some shots. I suspect the real Earl of Manchester had rather more to worry about in the English Civil War.

There have been many epics based on the military career of Napoleon Bonaparte, but in the western cinema, the best has to be Sergei Bondarchuk and Dino de Laurentis' *Waterloo* (1969). The battle itself was filmed near Uzhgorod in the Ukraine and the armies (there were 30,000 under Wellington's command alone) were made up of 16,000 extras from the Red Army, whose officers took their commands from Bondarchuk at the end of a walkie-talkie. At the time, *Waterloo* was one of the most expensive films ever made, racking up a £12 million bill. A full brigade of Soviet cavalry provided the horsemen for both sides – only the uniforms had changed! – and filming took forty-eight days. To make the Ukraine look like Belgium (fifteen years before Belgium was created) the Russians built 5 miles of roads, sowed barley, oats and flowers, planted 5,000 trees and erected Waterloo's key buildings of Hougoumont and La Haye Sainte.

Rod Steiger rides Napoleon's grey, Marengo (which bolted with the actor on his back), and his harness and saddle accoutrements are exact replicas of the original. Christopher Plummer sits Wellington's chestnut Copenhagen, although the bizarre scene where the horse lashes out with its hind hoofs when Wellington dismounts after sixteen hours in the saddle is not in the screenplay. It did not help that neither Steiger nor Michael Wilding, who plays General Ponsonby, the British cavalry commander, could ride well enough for the gallop. The result is an artificial series of shots in which both actors *pretend* to be riding – and it shows.

There are two cavalry 'moments' in *Waterloo*, both of them reasonably faithfully recorded. The first was the charge of the Scots Greys, the British heavy cavalry immortalised by Lady Butler in her painting *Scotland Forever.* As we have seen, Wellington complained that his cavalry in the Peninsula got him into scrapes and the same was true at Waterloo. The Greys charged, while at the walk, with Highland infantry clinging to their stirrup leathers, against a French formation that was too strong for them. Despite the stalwart work of Sergeant Ewart in capturing a French eagle (not shown in the film) the Greys exhausted themselves and were subsequently cut up by the fresher French lancers. Steiger delivers the famous line about the Greys – 'those men on grey horses are terrifying' – and one of his aides gets the balance right when he says, 'They are the finest cavalry in Europe and the worst led.' The horses and the uniforms are impressive but, as always, it is the stills in the publicity photographs that let the whole thing down. The bridles are wrong and at least half of the riders are carrying curved, as opposed to straight, swords.

The extras for the Greys were the Moscow Militia, interspersed with Yugoslav stuntmen and the charge, filmed over several days, was captured by Panavision cameras on trucks, via an aerial helicopter angle and the use of a train moving alongside the cavalry lines. Where the sequence failed was in the sudden use of slow motion. This was successful in *Braveheart* but the whole point of the Greys' charge was that it was delivered at the gallop – slowing the whole thing down merely exposed the wrong-patterned swords.

The second cavalry episode featured the mistake made by Marshal Michel Ney, 'the bravest of the brave' (played by a flame-haired Dan O'Herlihy). Ney misread Wellington's pulling back out of artillery range for a full-scale retreat and ordered his entire cavalry forward to complete what he thought was a rout. In fact, the British formed hollow square and the French horsemen rode round and round in useless circuits. Aerial helicopter shots show British squares breaking up all over the place – in reality, none of them did – and publicity stills show the infantry firing modern bolt-action rifles rather than the Brown Bess flintlocks they actually used. It is a credit to the film's production team that the torrential rain on the night before the battle was created by 6 miles of underground irrigation pipes. Most of the time, everybody sweated in temperatures of 95°F.

Probably the most celebrated cavalry disaster captured on celluloid was the *Charge of the Light Brigade.* Several versions have been made but only two stand out, both of them noteworthy in different ways. Errol Flynn's 1936 epic has gone down in history for the wrong reasons – the infamous use of tripwires to bring down galloping horses caused outrage at the time and since, not least from Flynn himself, who put in an official complaint about it. Appalling though this directorial device was, the alternative – of seeing horses miraculously get up after a fall while their rider lies 'dead' – is ridiculous. Infantry in particular aimed at horses – they were a bigger target and their collapse in an advance always caused havoc. In the movies, however, the loss of horseflesh is apparently non-existent. The storyline is pure hokum, involving the massacre of the British garrison at Cawnpore (an event that took place three years *after* the charge of the Light Brigade); an Indian prince/baddie sneakily in league with the Russians in the Crimea; and a regiment (Flynn's 27th Lancers) which did not exist until the Second World War. The actual charge is exciting, shot, complete with Tennyson's epic poem interspersed and, unusually, stills from the film show surprisingly accurate Lancer uniforms. Poor old Lord Cardigan, however, who actually led the charge, barely gets a mention, because it is Flynn himself who causes the charge by ignoring orders.

The 1968 remake is altogether more historical. The advisers for this version were the Mollo brothers, who set up a production company to research contemporary uniforms in detail. It is one of life's minor tragedies that the director, Tony Richardson, chose to ignore their meticulous work for reasons of cost or indifference. So the entire Light Brigade, not just Cardigan's 11th Hussars, wear crimson overalls. The Heavy Brigade (whose charge was filmed but cut from the final version) wear blue tunics, rather than scarlet, and so on. If the uniforms as envisaged by the Mollos were correct (which they were) the horses were even less well catered for. While Cardigan's Ronald has the accurate x-shaped bridle straps across the forehead, no one else seems to have acquired that and the sheepskin covers over the saddles are all of the wrong shape.

As with the tragedy of Roy Kinnear in the *Musketeers*, hardly anyone escaped uninjured during the making of this film. A stunt rider doubling for Flynn in the 1936 film was killed; *all* the stars in Tony Richardson's film fell off at least once, Mark Burns (playing Captain William Morris of the 17th Lancers) breaking his wrist. One thousand horses were used, along with 6,000lb of TNT and 3,000 cannonballs. The location in Turkey (as opposed to Flynn's California) is not bad for the Valley of Death at Balaclava, even if Captain Nolan's famous scarp slope down which he thundered carrying the fatal order that led to disaster, was far steeper in the film than in reality.

The only horse named in the film – other than Fanny Duberly's pony Bob – is Nolan's grey, Old Treasurer. In reality, this was Morris's horse and it was not grey. At Balaclava, Nolan rode a troop horse borrowed from the 13th Light Dragoons when he carried the order from Lord Raglan.

George MacDonald Fraser laments the fact that 90 per cent of American historical movies are westerns including cowboys in various settings. Here, we will confine ourselves to the American army and its cavalry in particular. Surprisingly few movies have covered the lives of Civil War legends like Robert E. Lee, 'Stonewall' Jackson, Ulysses Grant and 'Jeb' Stuart; they usually appear as bit-players in someone else's story. Lee, it is true, is central to the brilliant *Gettysburg* (1993, Ron Maxwell) but the central *military* action there is Pickett's Charge, delivered by infantrymen.

The Horse Soldiers (1959, John Ford) is a fictional account of a Union cavalry raid on a Southern railway junction vital to the Confederate war effort. The music is authentic, based on a contemporary love song, *Lorena*, and John Wayne is, as always, John Wayne. The appearance of earnest little boys led by their geriatric military academy teacher really did happen; Wayne's horsemen ride away rather than shoot them down. As such, *The Horse Soldiers* is pretty standard fare replayed scores of times in Indian war movies.

Major Dundee (1965, Sam Peckinpah) is refreshingly different, if only because it has a Civil War theme, an Indian element and a surprising clash with French cavalry. Charlton Heston, as Dundee, is in charge of a prisoner-of-war camp and takes some of them with his troopers to track down renegade Apache who have gone on the warpath. He has, he tells his men, 'but one order of march: if I say "Come" you come; if I say "Go" you go; if I say "Run" you follow me and run like hell.' All the tensions of the day are well handled – the rancour between Richard Harris's Southerners and Heston's coloured contingent; the mutual hatred for the French who are in Mexico to keep the Emperor Maximilian on the throne and the total contempt that everybody has for the Apache. Disappointingly, there is no obvious difference between French cavalry tactics and those of the Americans; perhaps because, by the 1860s, there was none.

For well over fifty years, the standard take in Indian westerns was, to quote General Phil Sheridan, 'the only good Injun is a dead Injun'.

Those of us who grew up watching these films at 'the pictures' had no concept of the bias, racism and inequality of any of this. *Of course* the Indians were the baddies. They could not even shoot straight, merely galloping hell-for-leather against a fort, cavalry formation or wagon train and riding round in circles, making a lot of noise and dust.

The John Ford series of westerns, filmed in the stark landscape of Utah's Death Valley, are quintessential examples of this but they also give us a surprisingly authentic peek at everyday life in the American cavalry. The best, for me, is *She Wore a Yellow Ribbon* (1949) about a cavalry colonel (John Wayne again) facing a retirement he does not want. He is surprisingly (for Wayne) sympathetic to the Indians but the hero of the film is the long-serving trooper, sitting his horse – or walking it, as was correctly done – through sweltering heat or grim, snowy winters. The horsemanship of Sergeant Tyree, played by real ex-cowboy Ben Johnson, is a joy to behold, even if the tactics of the army he worked for were rather ill-conceived.

Which brings us to perhaps the best-known version of several movies that have covered the Custer 'massacre'. *They Died With Their Boots On* (1941, Raoul Walsh) is a terrible title for a surprisingly good film, even if the history is very wide of the mark. Errol Flynn is George Armstrong Custer, portrayed as a well-meaning hothead rather than the vainglorious idiot he was. The Irish officer who taught Custer's 7th Cavalry the *Garryowen*, which became the regiment's anthem, was probably a trooper and it is unlikely that it happened anyway. Flynn's Custer has a sense of doom towards the end as he rides out to the Little Big Horn on a large chestnut horse that could either be Big Dandy or Vic (the two animals he owned). Interestingly, the 7th *do* dismount (as they did in reality) to try to outshoot the huge number of Lakota facing them. For cavalry to fight as dragoons in such a tight situation makes very little sense in the movie, as it probably did in 1876. We are also treated to Custer's repeated cavalry charges at Gettysburg when leading his Michigan cavalry, the Wolverines, during the Civil War. He rode five horses then, Lancer being his favourite.

There was a tentative American connection in *Young Winston* (1972, Richard Attenborough) if only because Churchill's mother, Jenny Jerome, was an American socialite. Churchill, regarded as a failure as a boy by his father, attended Sandhurst and was commissioned into the 4th Hussars in 1895. In the film, there is a delightful – and probably accurate – scene in which Simon Ward, playing the future prime minister – comes a cropper in the riding school. The Roughrider Sergeant instructing him shakes his head ruefully – 'You'll never get to India riding like that, Mr Churchill!'

The lieutenant's first posting was against the Pathan tribesmen on the Indian North-West Frontier and Churchill bought himself a grey horse. As one of his fellow officers asks a comrade in the film – 'Who's the bloody fool on the grey?' White is the most conspicuous colour on a battlefield and Churchill is taking a huge gamble by riding such a mount. By a series of coincidences and chances, Churchill found himself attached to the 21st Lancers in the Sudan expedition of 1898 and so rode one of the last cavalry charges in history. It is well done in the film, Ward swinging his grey (a different horse by this time) into line at the head of his troop. Technically, as a lieutenant, he should have ridden serrefile, *behind* his unit, but this would have looked less heroic on the screen. Seeing a mass of 'fuzzy-wuzzies' (Dervishes) rising up out of a concealed ditch, Ward mutters 'bloody hell', sheathes his sword and draws his Broomhandle Mauser pistol instead. Churchill actually did this, but only because an old polo injury was giving his arm gyp on the day. Location filming took place in the Sudan, but Morocco filled in for

India. Temperatures soared to 115°F in the shade and the problems of supply and accommodation were not dissimilar to those of the real British army in the days of empire.

Churchill's brilliant political career was over by 1917 and the future prime minister was a colonel in the trenches of the Western Front. Three thousand miles away, in Palestine, the Australian Light Horse were forming up to make their madcap dash for Beersheba. The Australians, along with the New Zealanders, had been 'ambushed' by the British high command in the futile Gallipoli campaign two years earlier but they were still there, fighting for General Allenby in the Middle East. And the guidon they carried in *The Light Horsemen* was not the Australian flag but the Union Jack.

The film was made in 1987 (Simon Wincer), based on an earlier effort *The 40,000 Horsemen* (1940), directed by Charles Chauvel, who was the nephew of Sir Harry Chauvel who had commanded the division in 1917. The attack on Turkish-held Beersheba was regarded as a suicide mission but the Australians accomplished it with panache. Technically, the Light Horse were mounted infantry, riding to battle and fighting on foot as the dragoons of the seventeenth century had, but that day, they charged as full-blown cavalry. The tactics, harness and weaponry of the unit was spot on. One critic said, 'I haven't seen a better action scene with horses since *Ben-Hur*' (which showed a chariot race, not a battle). The *Washington Post* was more dismissive – 'Mostly ... equine cinematography, a four-legged coffee table movie about the Australian cavalry.' Perhaps the critic was miffed because there have been so many bad movies about the American cavalry!

In 1982, Michael Morpurgo wrote his children's novel *War Horse*, having met old men who had served in the First World War with the Devon Yeomanry. Director Steven Spielberg had never worked with horses before – 'I was really amazed at how expressive horses are and how much they can show what they're feeling.' Shot in England over sixty-three days, the film was perhaps unique in that it showed turbaned Indian cavalry regiments in France (totally ignored by other film makers). The climactic charge, led by Benedict Cumberbatch on Joey, the eponymous horse-hero, was filmed at Strathfield Saye in Hampshire, the country seat of the Duke of Wellington and the burial place of his horse, Copenhagen.

The animals (300 of them) were trained over months by Bobby Lovegren and his team, 280 of them being used for the charge. The horses even had their own makeup artists to create 'wounds' and to ensure continuity. The United States Humane Association was on hand throughout and the film was lauded for 'outstanding' care taken of the animals. In the scene where Joey (one of fourteen different animals used) is tangled in barbed wire, the wire is actually rubber or the horse is animatronic. The thumping music score was written by John Williams who spent days at a California stud to prepare the work. 'Horses,' he wrote, 'are very special creatures. They have been magnificent and trusted friends for such a long time and have done so much for us with such grace.'

The storyline of *In Pursuit of Honor* (1995, Ken Olin) never happened, but the film is perhaps the most poignant salute to the warhorse ever made. Set in 1932, it involves an officer and a handful of troopers of the 12th US Cavalry 'stealing' horses rather than slaughtering them under the orders of General Douglas MacArthur as the American army becomes mechanised.

The film was shot in Australia, despite its Montana/Dakota setting. There is no record of any such slaughter taking place; in fact, horses were used throughout the 1940s by US military authorities and even into the Korean War (see Sergeant Reckless, Epilogue).

The most brilliant scene is when the deserters are making a run for the Canadian border, driving their doomed animals ahead of them. The bugler sounds the advance and 200 heads come up, ears upright, nostrils wide. Magic stuff!

'They fight beside us,' says a soldier in the film, 'and they die beside us – they deserve the same respect.'

Epilogue
'Goodbye, Old Man'

At 4.45 ack emma, on 1 September 1939, fifty-three German divisions, including six armoured units and motor transport invaded Poland. There had been no declaration of war, which was the customary protocol and the reason for invasion had been carefully engineered by the Germans themselves. Unbeknownst to millions involved, Adolf Hitler and Josef Stalin, rulers of Nazi Germany and Soviet Russia, had agreed to partition Poland between them and *Fall Weiss* was the codename for the German part of the plan.

Many of the older textbooks today refer to a futile cavalry action (of which film footage exists) showing the Polish horsemen charging the German panzers with lances. This never happened. Certainly the Poles were outgunned in every respect and their armoured division was particularly weak, but horses versus tanks was pure propaganda.

Horses were used extensively in the Second World War. Unlike the First, it was a mobile war, far more wide-ranging than 1914–18, relying on the vastly improved tanks in epic land battles like El Alamein and on aerial warfare using fighters, bombers and, perhaps ultimately, the atom bomb. But some parts of the world did not lend themselves to modern warfare of the machine age. Pack horses and mules were vital in the dense jungles of Malaya. At Bataan, American rations ran so low that men were reduced to eating their horses. Hitler's Panzers ground to a halt in Operation Barbarossa because of the severity of the Russian winter, but Stalin's Cossacks could still cover vast distances in the saddle. The USSR had thirty cavalry divisions, adding up to 1.2 million horses.

Ironically, in the months leading up to the war, Germany was buying up thousands of horses, for transport and reconnaissance at the same time that Britain, thanks to its policy of appeasement, was cutting back. In July–August 1940, the German army had 791,000 horses; America (still out of the war in its return to 'splendid isolation') had 9,000.

The last *actual* cavalry charge (as opposed to Omdurman and Moreuil Wood which both make that claim) took place in November 1941 near Musino, north-west of Moscow. The 106th German Infantry Division and the 107th Artillery suddenly found themselves under attack from Cossack horsemen, charging at the gallop as their ancestors had done for centuries. They were the 44th Mongolian Cavalry Division, armed with sabres and they were within 1,000 yards of the German line when the guns opened up on them. An estimated 2,000 horses and men were killed in ten minutes and only thirty reached the German guns. There were no German casualties.

But in every country involved in the Second World War, the writing had been on the wall for horsed cavalry for years. In Britain, the 11th Hussars, which had once thundered down the Valley of Death behind Lord Cardigan, became the first cavalry regiment to become mechanised; the 12th Lancers followed soon after. Men who had been used to a life of polishing buttons and sword blades and mucking out stables, now had to get used to squeezing into a tank or an armoured car with a whole range of new skills and problems. By the end of 1941 only two regiments, the Cheshire Yeomanry and the Yorkshire Dragoons, still had their horses and they both lost them within months.

Shortly after the Armistice was signed in 1918, an Italian artist, F. Matania, painted 'Goodbye, Old Man', a touching scene of a downed artillery horse. The caisson lies shattered in a ditch next to the road and a broken-hearted artilleryman is kissing the dying animal on the nose, his harness removed, his eyes rolling in his head. The picture was reproduced as a cheap print and sold in its thousands in the years immediately after the war.

The prime minister, Lloyd George, paid lip service to the cavalry of the First World War, especially Allenby's units, but he authorised the abandoning of some 20,000 horses to be sold off in Egypt. Protests from horse-lovers everywhere, even someone with the clout of Major General George Barrow, commanding the Yeomanry Division, achieved nothing. The war had cost a fortune; belts must be tightened and horses were, as they had always been, expendable.

The fate of these horses became the obsession of Dorothy Brooke, wife of the brigadier commanding the Cavalry Brigade in Egypt in 1930. Many officers shot their chargers in 1919 rather than let them end up at the Egyptians' tender mercies, but 20,000 were too many. Mrs Brooke wrote that the rest were 'always hungry, weak, over-loaded ... lame, crippled, galled, ill-shod, frequently blind, suffering from perpetual thirst ... tormented by flies ... straining under the whip.' Mrs Brooke found many horses still alive but in a terrible condition. They were recognisable by the arrow brand on their hindquarters. Buying as many as she could afford, Mrs Brooke enlisted the help of the London *Morning Post* and donations flooded in, at first to the tune of £600, then, four years later, £40,000. Five thousand horses were rescued but since several were, by now, over 20 years old, they had to be put down. The Brooke Hospital for Animals in Cairo continues its humane charity work to this day.

In June 1934, the London International Horse show staged a belated parade of veteran horses. After so many years since the end of their active service, only twenty-five could be found. The oldest was thirty-two and they were paraded every evening of the three-day event draped with shabraques embroidered with their battle honours. Warrior was there, as was Quicksilver, the mount of Colonel Percy Laine, assistant-commissioner of the Metropolitan Police. The only infantry charger to have gone through the war with one battalion and having never missed a day was Kitty, the horse of Lord Digby of the Coldstream Guards. Nigger (a name unthinkable today) had charged with the Buckinghamshire Hussars at El Mughar. David was a wheel-horse with the Royal Field Artillery; four officers clubbed together and bought his retirement in Hertfordshire in 1926. A complete gun team, the 'Old Blacks', took part in this parade, having served throughout the First World War with three of their original four drivers still alive. The animals had drawn the gun carriage at the funeral of the Unknown Warrior in Westminster Abbey.

There are a number of memorials to the cavalry dotted around the countries that fought the Great War. Many of those relate to individual regiments, like that of the 6th Dragoon Guards, the Carabiniers, dedicated to the men who served. When J.M. Brereton wrote *The Horse in War* in 1976, he lamented the fact that there was no memorial to the *horses*, as opposed to the men who rode them. That has now been rectified. The brilliant sculpture that forms the dustjacket image for this book stands in Park Lane, just outside Hyde Park. It portrays a horse and mule, doggedly ambling up to a gate, laden with baggage. Beyond the gate, the baggage has gone and the horse is galloping free. This is a tribute to all animals who served king and country. The memorial purely for the 1914–18 warhorse stands at Ascot, Berkshire, long linked with racing. It is a larger than life bronze cavalry horse on a three metre tall stone plinth and, appropriately, barbed wire coils between its hoofs. It was unveiled on 8 June 2018 after a number of memorials and poppy displays appeared all over Britain and Europe. The horse, too, is called Poppy.

The Korean War (1950–53) is an unlikely setting for a heroic horse, yet Sergeant Reckless fits the bill. Korea was a 'hot point' in the Cold War that developed between East and West after the defeat of Nazi Germany. Looked at now, it seems an odd halfway house between the Second World War and the American shambles of Vietnam. Military technology was improving all the time, with ever longer range guns and the arrival of the jet aircraft and the helicopter. After the dropping of the atom bombs on Hiroshima and Nagasaki in 1945, the world held its breath that such a devastating weapon would never be used again.

In that sense, the use of Reckless was a steadying influence, a symbol of past wars, giving a perception of safety. The chestnut mare with a white blaze and three white socks was probably 4 when the US Marine Corps bought her. Her original name was Ah Chim Hai, which means Sun in the Morning and she was a 14 hands Mongolian with Thoroughbred strains. Her duties were to carry 24lb shells for the Recoilless Rifle Platoon of the 5th Marine Regiment. Hence her new name – Recoilless became Reckless. She carried up to eight shells at a time and by the time of the battle of Outpost Vegas in March 1953, travelled a total of 35 miles on her own, under heavy fire throughout. She was hit by shrapnel twice. The Marines taught her to lie down when the firing started and to avoid barbed wire.

Reckless was adored by her unit. She drank two bottles of Coke a day, ate anything, including (twice) a horse blanket, bacon sandwiches, even poker chips. She slept near the stove in winter and had the run of the camp. For her courage under fire, she was promoted, first to corporal, then, after her retirement, to staff sergeant. On that occasion, she had a 1,700-man guard of honour and a nineteen-gun salute.

Her former commander, Lieutenant General Randolph M. Pate said, 'Like any other Marine, she was enjoying a bottle of beer with her comrades ... If she failed to receive the attention she felt her due, she would deliberately walk into a group of Marines and, in effect, enter the conversation.'

There can be no finer tribute to a soldier's mount than the words written by Ronald Duncan for the Horse of the Year Show in Britain in 1954, when the Korean War was over and a full century after the British Light and Heavy Brigades charged to glory:

> He serves without servility; he has
> fought without enmity. There is nothing so
> powerful, nothing less violent, there is nothing so quick,
> nothing more patient.
>
> England's past has been borne on
> his back. All our history is
> his industry; we are his heirs, he
> our inheritance.

And in the Great War itself, the poet Julian Grenfell said much the same thing in *Into Battle*:

> In dreamy, doubtful waiting hours
> Before the brazen frenzy starts,
> The horses show him nobler powers:
> O, patient eyes, courageous hearts!

Bibliography

Adkin, Mark, *The Waterloo Companion,* Aurum Press, 2001
Anglesey, Marquess of, *A History of British Cavalry 1816–1919 3 volumes,* Leo Cooper
Boniface, Col. John J., *The Cavalry Horse and Pack,* Must Have Books, Canada, (Reprint of 1903 edition)
Brereton, J.M., *The Horse in War,* David and Charles, 1976
Brighton, Terry, *The Last Charge,* The Growood Press, 1998
Brooks, Richard, *Cassell's Battlefields of Britain and Ireland,* Weidenfeld and Nicholson, 2005
Carr-Gomm, Philip and Stephanie, *The Druid Animal Oracle,* Grange Books, 2001
Cavalry Training, War Office, London, 1904
Chappell, Mike, *British Cavalry Equipments,* Osprey, 1983
Cooper, Leonard, *British Regular Cavalry,* Chapman and Hall, 1965
Davies, John, *The Celts,* Cassell & Co, 2000
Dillon, Richard H., *North American Indian Wars,* Magna Books, 1983
Edge, David and Paddock, Miles, *Arms and Armour of the Medieval Knight,* Guild Publishing, 1988
Embleton, Gerry and Howe, John, *The Medieval Soldier,* Windrow and Greene, 1994
Glover, Michael, *Warfare in the Age of Bonaparte,* BCA 1980
Goldsworthy, Adrian, *Roman Warfare,* Cassell, 2000
Goldsworthy, Adrian, *The Complete Roman Army,* Thames and Hudson, 2003
Gravett, Christopher, *Knights at Tournament,* Osprey, 1988
Guest, Ken and Denise, *British Battles,* Harper Collins, 2002

Hamilton, Jill, *Marengo; the Myth of Napoleon's Horse,* Fourth Estate, 2001
Haythornthwaite, Philip J., *The Colonial Wars Source Book,* BCA 1995
Haythornthwaite, Philip J., *The Napoleonic Source Book,* Arms and Armour Press 1990
Holmes, Richard, *The Napoleonic Wars Experience,* Andre Deutsch, 2006
Hook, Jason, *The American Plains Indians,* Osprey, 1985
Keen, Maurice, *Chivalry,* Yale University Press, 1984
Kegan, John and Wheatcroft, Andrew, *Who's Who in Military History,* BCA, 1976
Knight, Ian, *Go To Your God Like a Soldier,* Greenhill Books, 1996
Lawford, James, (ed), *The Cavalry,* Sampson Low, 1976
Marshall, Robert, *The Storm from the East,* BCA, 1993
Matthews, John and Stewart, Bob, *Celtic Battle Heroes,* Firebird Books, 1988
Matyszak, Philip, *The Enemies of Rome,* Thames and Hudson, 2004
Mollo, Boris, *The Indian Army,* Blandford Press, 1981
Mortimer, Ian, *The Time Traveller's Guide to Elizabethan England,* Bodley Head, 2012
Mortimer, Ian, *The Time Traveller's Guide to Medieval England,* Vintage Books, 2009
Musset, Lucien, *The Bayeux Tapestry,* Boydell Press, 2002
Newark, Peter, *Sabre and Lance,* Blandford Press, 1987
Nicolle, David, *Mughul India 1504–1761,* Osprey, 1993
Nolan, Captain L.E., *Cavalry: Its History and Tactics,* London, 1854
'Notrofe', *Cavalry Taught by Experience,* Hugh Rees Ltd, 1910
Peddie, John, *The Roman War Machine,* Sutton Publishing, 1994
Peterson, Daniel, *The Roman Legions Re-enacted in Colour Photographs,* Crowood Press, 1992
Purkiss, Diane, *The English Civil War,* Harper Press, 2006
Reedstrom, E. Lisle, *Custer's 7th Cavalry,* Sterling Publishing, 1992

Ross, Anne, *Pagan Celtic Britain*, Constable, 1967
Sekunda, Nick, Northwood, Simon and Simkins, Michael, *Caesar's Legions*, Osprey, 2001
Simkins, Michael, *The Roman Army from Caesar to Trajan*, Osprey, 1984
The Military Horse, Marshall Cavendish, 1976
Trow, M.J., *El Cid: the Making of a Legend*, Sutton Publishing, 2007
Turnbull, S.R., *Samurai Armies 1550–1615*, Osprey, 1979
Turnbull, Stephen, *The Book of the Medieval Knight*, Arms and Armour Press, 1985
Tylden, Major G., *Horses and Saddlery*, J.A. Allen, 1965
Vinksie, V., and Grbasic, Z., *Cavalry*, Cassell, 1993
Webber, Christopher, *The Thracians*, Osprey, 2000
Wilcox, Peter, *Rome's Enemies – Gallic and British Celts*, Osprey, 1985
Wrangel, Alexis, *The End of Chivalry*, Hippocrene Books, 1982
Young, Peter and Adair, John, *Hastings to Culloden*, G. Bell & Sons, 1964

Index

Alans 52
Alexander the Great xv, 9-20, 209
Alexander, Pte John 153-5
Allenby, General Edmund 204-206, 223, 228
Ancient Egyptians 5, 58, 187, 208
Apache 220
Arapaho xiii, 174
Assyrians 2, 4-5, 57
Avars 52-3

Baker, Colonel Valentine 146, 181-2
Battles:
  Aliwal 179
  Alma 148, 151
  Austerlitz 125
  Balaclava xvi, 40, 219
  Bannockburn 56, 63-4, 88
  Blenheim 113
  Bosworth 87-8, 214
  Bull Run 158, 162
  Chaeronea 12-16
  Crecy 79-81, 84
  Dettingen 118
  Eylau 125
  Falkirk 213
  Ferozeshah 179
  Friedland 125
  Gaugemala 16, 18
  Granicus 16-18
  Hastings 55-8, 60, 66, 79, 83, 88
  Issus 14
  Jena 110, 125, 127, 136
  Ligny 132
  Little Bighorn xii, 165, 169, 172-6
  Malplaquet 113
  Manassas – see Bull Run
  Manduessedum 30
  Naseby 98, 100, 103-104
  Omdurman 187-9, 227
  Oudenarde 113
  Quatre Bras 132
  Ramillies 113
  Sobraon 179
  Ulm 125
  Wounded Knee 175
Benteen, Captain Frederick xii, 166, 169
Black Beauty 179
Blackfoot 173
Bonaparte, Napoleon 110, 114, 121, 123-9, 132, 135-7, 142, 158, 160, 185, 200, 204, 209, 216
Boudicca 29-30, 38, 45
Brooke Hospital for Animals 228
Brooke, Mrs Dorothy 228
Brudenell, James Earl of Cardigan 139-43, 146, 148, 150-2, 218-19, 227
Buffalo Calf Road Woman 174

Caesar, Julius 21-3, 25, 30, 34, 47
Caligula 32-7, 43, 47
Cantiaci 21-4
Cavendish, William Duke of Newcastle 113
Centaurs 18-19, 92
Chauvel, General Henry 206
Cheyenne xiii, 166, 168, 171, 173-5
Churchill, John Duke of Marlborough 105, 111, 113-14
Churchill, Winston 182, 188, 191, 198, 205, 222-3
Clarke, Colonel William xvi, 149
Clarke, Lt W.P. 175
Cortes, Hernan 91-6
Cossacks 40, 111, 117, 148-9, 198, 215, 226-7
Cox, Staff Sgt Jack 193-6
Cree 173
Cromwell, Oliver 12, 98, 101-106

Crow 170-1, 174
Custer, General George Armstrong xii-xiv, 160, 165-6, 169-73, 175-7, 221-2

de Narvaez, Panfilo 92
DeLacey, Sgt Milton xii-xiii, 176
Dervishes 188-92, 222
Diaz, Rodrigo of Bivar – see El Cid

El Cid 60-3, 211
Elliot, Lt Alexander 149
Epona 25-6, 28

Fairfax, Thomas 89, 99, 101, 105
Fane, Colonel Walter 180
Films:
   Cromwell 215
   Alexander 209
   Alexander Nevsky 211
   Braveheart 212-13, 217
   Charge of the Light Brigade (1936) 218
   Charge of the Light Brigade (1968) 219
   El Cid 211
   Genghis Khan 212
   Henry V 213-14
   Major Dundee 220
   Pursuit of Honor xi, 224-5
   Richard III 214
   Richard 214
   She Wore a Yellow Ribbon 221
   Spartacus 40, 209-10
   Taras Bulba 215
   The Horse Soldiers 220
   The Lighthorsemen 223
   They Died With Their Boots On 221-2
   Warhorse 224
   Waterloo 216-17
   Young Winston 222-3
Frederick the Great 109-10, 116-17

Gaius Julius Caesar Germanicus – see Caligula
Genghis Khan xvi, 75-8, 212
Godman, Lt Richard 17, 146, 149
Godwinson, Harold 56-8
Grant, General Ulysses S. 163-4, 220
Gustavus Adolphus 103-104

Hengest 48-50
Hittites 3
Hodge, Lt Col Edward 145, 147-8, 150, 182
Hodson, Colonel William 180
Horsa 48-9
Horses: Breeds and Types
   Andalusian 60-1, 92, 115
   Appaloosa 157
   Arab 64-6, 78, 87-8, 91, 107, 110, 123, 129, 147, 181-2, 187, 198
   Barb 87, 106-107, 114, 137, 181
   Cape 181
   Destrier 58, 61, 63-4, 68-9, 79, 81-3, 85, 213
   Eohippus xii-xvii
   Fjord 210
   Friesian 114
   Holsteiner 115
   Mesohippus xiv, xv
   Mongolian xvi, 72-4, 76-8, 230
   Morgan 157
   Mustang 157, 184
   Nisean 6
   Oldenburg 114
   Palfrey 58, 68-9, 82-3
   Przewalski xvi, 73
   Quarterhorse 157, 184, 198, 212
   Rouncey 58, 68
   Saddlebred 157, 161
   Standardbred 157
   Tarpan xvi
   Thracian 8
   Waler 181, 187, 194-5, 198
Horses:
   Ali 129
   Babieca 55, 60-2
   Big Dandy 221
   Black Barbarie 97-100, 215
   Bobtail *see* La Rabona
   Bucephalus xv, 10-20, 208
   Byerly Turk 110
   Cadet 130
   Cedric ix
   Cincinnati 164
   Comanche xiii, 167, 176-7
   Copenhagen 119, 121, 137-8, 216, 224

Darley Arabian 110
David 229
Drover *see* Tziminchak
Drummer Boy 150
Echo x
El Arriero *see* Tziminchak
Epaulette ix
Exquisite 150-1
Falcon ix
Gonsalvo 129
Incitatus 32-7, 46-7
Intendant (Coco) 130
Kitty 229
La Rabona 93
Lyard Gilder 83
Marengo 129-30, 136-7, 216
Montevideo 129
Moron 93
Morzillo *see* Tziminchak
Moses 150
Nigger 229
Old Treasurer 150, 219
Pegasus 18
Poppy 230
Punch 181
Quicksilver 229
Rainbow 190
Roan Barbary 86-8
Rochester ix
Roitelet 129
Ronald 140-3, 150-2, 219
Sefton viii-xi
Sergeant Reckless 230
Sorrell Blackwell 83
Spitfire 150
Sultan xvi, 149
Tauris 129
The Earl 148-9
Traveller 161-3
Tziminchak 91, 93, 95-6
Vic 165, 221
Volonel 183
Wagram (Ingenue) 129
Warrior 203-204, 229
Waterford ix
White Surrey 87-9, 214
Zara ix

Huns xv, 44, 50-2

Ireton, Henry 98-101

Keogh, Captain Myles xiii, 165-7, 169, 175-6
Kikkuli 3

Lakota xiii, 165-6, 171, 173-4, 221
Lasalle, General Antoine 127
La Marchant, Colonel John 131
Lee, General Robert E. 159, 161-3, 169, 220

Macro, Quintus Sutorius 32-4, 36
Mamelukes 123, 127
Marsh, Othniel C. xiv
Marshal, William 68
Martin, Colonel Rowland 187-91
Mitani 2-3
Mongols 42, 68, 70-8, 212
Montezuma 94
Morris, Captain William 140-1, 144, 150-1, 219
Mosby, Colonel John 155, 160
Mouat, Surgeon James 151
Murat, Joachim 40, 122, 135-6

New Model Army 89, 98, 103, 105-106, 108
Ney, Marshal Michel 217, 134-5
Nolan, Captain Louis 3, 139-41, 144, 146-7, 189, 219

Odo, Bishop of Bayeux 55, 57-8
Osborn, Henry F. xiv

Parthians 4, 44, 52-3, 67, 74
Pedersen, Trooper Michael viii-x
Philip of Macedon 10-12, 16
Poniatowski, Marshal Josef 127
Ponsonby, General William 120, 133-4
Probyn, Colonel Dighton 180
Przewalski, Captain Nicolai xvi, 73

Quantrill, William 161

238 Famous Horses at War

Regiments:
    4th Australian Light Horse 193, 195, 206
    4th Dragoon Guards 182
    4th Light Dragoons 140
    5th Dragoon Guards 17, 146
    5th Lancers 204
    7th Hussars 146
    7th US Cavalry xii–xiii, 169-70, 172, 175-6, 221
    9th Lancers 182, 186
    10th Prince of Wales's Own Hussars 182
    11th Prince Albert's Own Hussars 140, 143, 219, 227
    12th Australian Light Horse 206
    12th Lancers 182, 227
    12th US Cavalry 224
    13th Light Dragoons 148, 219
    15th Hussars (Light Dragoons) 115
    16th Lancers (Light Dragoons) 184
    17th Lancers 140-1, 143, 184, 187, 220
    21st Lancers 187-92, 222
    43rd Virginia Cavalry 154-5
    Army Remount Department 198
    Buffalo Soldiers (9th and 10th US Cavalry) 156, 168-9
    Carabiniers (6th Dragoon Guards) 229
    Cherokee Mounted Rifles 156
    Fort Garry Horse 203
    Hampshire Yeomanry 195, 202
    Horse Artillery 117, 126
    Inniskillings (6th Dragoons) 147, 204
    King's German Legion 131
    Life Guards 40, 110, 130, 142
    Lord Strathcona's Horse 203-204
    Queen's Bays (2nd Dragoon Guards) 110
    Royal Canadian Dragoons 203
    Royal Dragoons (1st) 150
    Rush's Lancers (6th Pennsylvania Cavalry) 157
    Scots Greys (2nd Dragoons) xvi, 46, 1110, 130, 133, 149
    Tarleton's Light Dragoons 184
    X Gemina Legion 21-22
Reno, Major Marcus xii, 166, 169
Richard I 63-4, 68, 84
Richard II 84, 86-8, 106
Richard III 87-109, 87
Robert the Bruce 56, 69
Roberts, General Frederick 182, 186, 197
Robertson, Field Marshal William 178
Rudolf, Count of Lorraine 79-81, 84
Rupert, Prince 97-100, 104-105, 113, 117, 130, 215

Saladin 64, 68
Sassenids 7
Saxe, Marshal (Maurice of Saxony) 115
Scarlett, General James 139, 148-9
Scythians xv, 6-7, 52
Seely, Colonel Jack 202-204
Sherman, General William T. 159, 164
Shoshone xiii
Somerset, Fitzroy Lord Raglan 139-40, 148-50, 152, 219
Spartacus 210
Stand Watie, General 156
Stewart, General 'Jeb' 157-60, 220
Subedei B'atatur 70-2, 74-7

Tarleton, Banastre 184
Temujin *see* Genghis Khan

Waterloo 46, 119-22, 126, 129-30, 133, 136, 142, 199, 217
Wellesley, Arthur Duke of Wellington 119-24, 130, 132, 134, 136-8, 144, 146, 187, 199, 216-17, 224
William I 55-60, 66, 106

Xenophon 6, 8, 13-15

Yorke, Col John 150-1